Synthesebücher

SCHWERPUNKTPROGRAMM **UMWELT**
SCHWEIZ. NATIONALFONDS ZUR FÖRDERUNG DER WISSENSCHAFTLICHEN FORSCHUNG
PROGRAMME PRIORITAIRE **ENVIRONNEMENT**
FONDS NATIONAL SUISSE DE LA RECHERCHE SCIENTIFIQUE
PRIORITY PROGRAMME **ENVIRONMENT**
SWISS NATIONAL SCIENCE FOUNDATION

Transdisciplinarity: Joint Problem Solving among Science, Technology, and Society

An Effective Way for Managing Complexity

J. Thompson Klein
W. Grossenbacher-Mansuy
R. Häberli
A. Bill
R. W. Scholz
M. Welti
(eds.)

Springer Basel AG

Editors

Julie Thompson Klein
Interdisciplinary Studies Program
Wayne State University
Detroit, MI 48303, USA
e-mail jklein4295@aol com

Rudolf Haberli
Swiss Priority Program Environment
Swiss National Science Foundation
Langgassstrasse 23
CH-3012 Bern, Switzerland
e-mail haeberli@sppe ch

Roland W Scholz
Swiss Federal Institute of Technology (ETH)
Chair of Environmental Sciences –
Natural and Social Science Interface
Haldenbachstrasse 44
CH-8092 Zurich, Switzerland
e-mail scholz@uns umnw ethz ch

Walter Grossenbacher-Mansuy
Swiss Priority Program Environment
Swiss National Science Foundation
Langgassstrasse 23
CH-3012 Bern, Switzerland
e-mail grossenbacher@sppe ch

Alain Bill
ALSTOM Power Technology Ltd
Segelhof 1
CH-5405 Baden-Dattwil, Switzerland
e-mail alain bill@power alstom com

Myrtha Welti
Swiss Foundation Science et Cité
Marktgasse 50
CH-3011 Bern, Switzerland
e-mail myrtha welti@science-et-cite ch

A CIP catalogue record for this book is available from the Library of Congress, Washington D C , USA

Deutsche Bibliothek Cataloging-in-Publication Data
Transdisciplinary joint problem solving among science, technology, and society an effective way for managing complexity / ed by Julie Thompson Klein (eds) – Basel , Boston , Berlin Birkhauser, 2001
(Synthesebucher Schwerpunktprogramm Umwelt)

ISBN 978-3-0348-9559-0 ISBN 978-3-0348-8419-8 (eBook)
DOI 10.1007/978-3-0348-8419-8

Originally published by Birkhäuser Verlag in 2001
Softcover reprint of the hardcover 1st edition 2001

Printed on acid-free paper produced from chlorine-free pulp TCF ∞
Cover design Sandra Baumann, Bern

9 8 7 6 5 4 3 2 1

Contents

Text References

Throughout this volume, Julie Thompson Klein presents abstracts at the head of each chapter. References to this volume and the two pre-conference Workbooks are italicized in chapters; full references to Workbook texts, by authors/names, appear in Appendix D: List of Contributors. Workbook texts are available at <http://www.transdisciplinarity.ch>. The URL is maintained and updated by SAGUFNET.

Transdisciplinarity: Joint Problem-Solving among Science, Technology and Society. Proceedings of the International Transdisciplinarity 2000 Conference (Zurich: Haffmans Sachbuch Verlag, 2000).

Workbook I: Dialogue Sessions and Idea Market, ed. R. Häberli et al.
Workbook II: Mutual Learning Sessions, ed. R. Scholz et al.

Foreword

What kind of science do we need today and tomorrow? In a game that knows no boundaries, a game that contaminates science, democracy and the market economy, how can we distinguish true needs from simple whims of fashion? How can we distinguish between necessity and fancy? How can we differentiate conviction from opinion? What is the meaning of this all? Where is the civilizing project? Where is the universal outlook of the minds that might be capable of counteracting the global reach of the market? Where is the common ground that links each of us to the other?

We need the kind of science that can live up to this need for universality, the kind of science that can answer these questions. We need a new kind of knowledge, a new awareness that can bring about the creative destruction of certainties. Old ideas, dogmas, and out-dated paradigms must be destroyed in order to build new knowledge of a type that is more socially robust, more scientifically reliable, stable and above all better able to express our needs, values and dreams. What is more, this new kind of knowledge, which will be challenged in turn by ideas yet to come, will prove its true worth by demonstrating its capacity to dialogue with these ideas and grow with them.

Why should disciplines be brought together? It is becoming increasingly clear that all three historical forces – the market economy, science and democracy – not only benefit from, but more importantly require, working across disciplines. First and foremost, the market economy requires polydisciplinarity. The complexity of today's world demands that any given participant be able to combine the responsibilities and talents of multiple roles; we are called upon to be a new breed of hybrid professionals, i.e. both economists and lawyers at the same time, as well as engineer-economist or doctor-historian. Second, science itself requires interdisciplinarity. How else can we study the effect of climatic disorder on genomics, if not by adopting a transdisciplinary approach? Finally, democracy requires transdisciplinarity. Indeed, it is a sine qua non condition underpinning the proper functioning of the agora in a complex world.

Charles Kleiber, State Secretary for Science,
Berne, Switzerland

1

Summary and Synthesis

1.1 Summary

by Rudolf Häberli, Swiss Priority Program Environment (SPPE), Berne, Chair of the Conference; Walter Grossenbacher-Mansuy, SPPE; Julie Thompson Klein, Interdisciplinary Studies Program, Wayne State University, Detroit, Michigan, USA

International Transdisciplinarity Conference
Zurich, Switzerland

Program Overview

- Sunday, February 27, 2000 Welcome

- Monday, February 28, 2000 Opening

- Plenary Sessions: Keynote Adresses (Kleiber, Colwell/Eisenstein, Gibbons/Nowotny), Panel I (Bill/Oetliker/Thompson Klein)
- Dialogue Sessions

- Tuesday, February 29, 2000 Mutual Learning Sessions

- Wednesday, March 1, 2000 Plenary Sessions
- Keynote Adresses (Ernst, Schneidewind, Scholz/Marks)
- Panel II (Hollaender/Leroy), Award, Panel III (Mey/Kapila)
- Conclusions, Closing

Nearly 800 participants from about 50 countries attended the conference. It was organized jointly by the Swiss National Science Foundation, Berne, Swiss Priority Program Environment; the Swiss Federal Institute of Technology, Zurich, Natural and Social Science Interface (ETH-UNS); and Asea Brown Bovery – ABB Corporate Research Ltd, Baden-Dättwil. It was also held under the patronage of UNESCO and in cooperation with the Swiss Foundation Science et Cité.

1. Why Transdisciplinarity?

- The core idea of transdisciplinarity is different academic disciplines working jointly with practitioners to solve a real-world problem. It can be applied in great variety of fields.
- Transdisciplinary research is an additional type within the spectrum of research and coexists with traditional monodisciplinary research.
- The science system is the primary knowledge system in society. Transdisciplinarity is a way of increasing its unrealized intellectual potential and, ultimately, its effectiveness.

2. How is Transdisciplinarity Done?

- Transdisciplinary projects are promising when they have clear goals and competent management to facilitate creativity and minimize friction among members of a team.
- Stakeholders must participate from the beginning and be kept interested and active over the entire course of a project.
- Mutual learning is the basic process of exchange, generation and integration of existing or newly-developing knowledge in different parts of science and society.

3. How is Transdisciplinarity Promoted?

In General

- The most important element is recognition of this form of joint research and learning by society and influential individuals.
- Every proposal for a research project, in order to produce reliable and socially robust knowledge, has to answer the question: "Where is the place of people in our knowledge?"
- A culture of transdisciplinary cooperation is needed, more than transdisciplinary professorships or institutes.

By Appropriate Institutional Structures

- Flexible organizations must be established, with new stimuli such as taskforces and transdisciplinarity labs.

- Transdisciplinary funding means cooperative funding and requires combing funds oriented toward innovation processes and problem solutions.
- Courageous Research and Development administration is needed to promote transdisciplinarity, not simply praise interdisciplinarity and still promote disciplinarity.

By Powerful Incentives

- To promote transdisciplinarity, new incentives are necessary for young researchers to move across boundaries.
- Evaluation procedures and criteria must be adapted to support transdisciplinary researchers.
- Transdisciplinary phases in professional careers should have positive value and effects in the scientific system.

4. What Future Developments are Anticipated?

- The keyword for the 21st century is sustainability. Transdisciplinarity is one of the major tools for reaching it.
- Least-cost science will be the dictate of the future, not only in view of scarce funding budgets.
- Because the societal impact of transdisciplinary research may only become visible years after completion of a project, long-term monitoring is needed.

5. What are the Most Urgent Tasks?

- Three efforts are of utmost importance:
  - clarifying theoretical concepts and their value,
  - building on the existing empirical evidence, and
  - designing appropriate criteria, practices, and guidelines.
- Developing transdisciplinary research and teaching takes time and commitment from academics, their institutions and their partners in practice. Everyone must be patient.
- The conference was a starting point, not a single event. Networks, institutional structures, and exchanges of knowledge must be created and sustained.

1.2 Synthesis

by Rudolf Häberli; Alain Bill, ABB Corporate Research Ltd, Baden-Dättwil; Walter Grossenbacher-Mansuy; Julie Thompson Klein; Roland W. Scholz, ETH, Zurich; Myrtha Welti, Swiss Foundation Science et Cité, Berne

Abstract

Several basic questions are answered at the outset. What is transdisciplinarity, and what kind of research is it? Why and where is it needed? What was learned about main goals of the conference: (1) developing transdisciplinary practice (clear goals, careful preparation, competent management, funding for cooperative work), (2) promoting transdisciplinary research (openness on the sides of academia and practice, platforms for contact and promotion, adequate forms of evaluation), (3) creating favorable institutional structures (education and training, funding structures, incentives).

- "Mutual learning between science and practice"
- "Getting socially robust knowledge"
- "Problem solving in a *ménage à trois*"
- "Tear down the barriers between sciences"

These were the headlines in media reports on the International Transdisciplinarity Conference, held in Zurich, Switzerland from February 27 to March 1, 2000. The short definition provided by the Chair of the conference, Rudolf Häberli, summed up the leading ideas: "Transdisciplinary research takes up concrete problems of society and works out solutions through cooperation between actors and scientists." The partners in the household of the *ménage à trois* that State Secretary of Science Charles Kleiber defined – science, democracy and market economy – must forge new alliances if we are to meet effectively the challenges of the 21st century, called "the century of complexity."

Several basic questions follow from this brief description.

Perspectives on the Conference

What is Transdisciplinarity?

Transdisciplinarity is a new form of learning and problem solving involving cooperation among different parts of society and academia in order to meet complex challenges of society. Transdisciplinary research starts from tangible, real-world problems. Solutions are devised in collaboration with multiple stakeholders. A practice-oriented approach, transdisciplinarity is not confined to a closed circle of scientific experts, professional journals and academic departments where knowledge is produced. Ideally, everyone who has something to say about a particular problem and is willing to participate can play a role. Through mutual learning, the knowledge of all participants is enhanced, including local knowledge, scientific knowledge, and the knowledges of concerned industries, businesses, and non-governmental organizations (NGO's). The sum of this knowledge will be greater than the knowledge of any single partner. In the process, the bias of each perspective will also be minimized (*Hurni; Pohl*).

Why the Conference?

Since 1992, the Swiss National Science Foundation has sponsored the Swiss Priority Program Environmental Technology and Environmental Science (SPPE). Because environmental problems typically involve more than one academic discipline, the program management promoted transdisciplinary research from the beginning. With the forthcoming end of the SPPE in the year 2001, the program's managers conceived the idea of presenting its results to a wider audience, in the framework of an International Transdisciplinarity Conference. Since the challenges of society extend beyond environmental problems, two additional Swiss and international institutions became co-organizers: the Natural and Social Science Interface of the Swiss Federal Institute of Technology in Zurich (ETH) and Asea Brown Boveri, ABB Corporate Research Ltd, in Baden-Dättwil (*Bill; Häberli et al.; Scheringer et al.; Scholz a,b;* Scholz et al. 2000, Organizing Committee Transdisciplinarity Conference 1999; Swiss National Science Foundation 1995; Program Management SPPE 1995).

Discussions among the co-sponsors and an international Conference Board resulted in three interrelated goals:

(1) Developing transdisciplinary practice
(2) Promoting transdisciplinary research
(3) Creating favorable institutional structures and powerful incentives for transdisciplinarity.

This Synthesis reports on how these goals were reached.

Why Transdisciplinarity?

For important challenges, such as sustainability, expertise is not restricted to academic circles alone. North-South relations, new forms of business and labor, and rearrangements in our environment are required. Other parts of society must be involved as well, including industry, business, public administration, and non-governmental organizations (NGO's). Three forces – complexity, high intensity of conflicts and "language barriers" – reinforce the necessity of transdisciplinarity. It is an essential tool for creating new insights that lead to new solutions and engage creative processes of mutual learning, not just a peripheral approach (*Becker; Egger and Jungmeier; Flury; Peterson and Schaltegger*).

Although "transdisciplinary" is a relatively new word, the concept of taking up concrete problems of society and working out solutions in cooperation with scientists and other actors has a long tradition (*Adi*). The emergence of the information society, with its attendant societal and economic changes, has fostered a new democratization of knowledge and involvement of industries in the production and management of knowledge (Gibbons et al. 1994). Transdisciplinarity has mainly been classified as oriented research, but it can be situated in a wide field between basic and applied research (Fig. 1). It does not replace traditional forms of research, such as "free" and "basic" (disciplinary) research. It is an additional and mainly demand-driven form of research that involves partners from outside academia.

In oriented research, efforts are directed at specific subject-areas. Its opposite is free research, which is not bound to any specific subject area and in which researchers are responsible for determining the content of their work. We also distinguish between applied research, which directly serves certain social purposes (the economy, politics, other areas of society) and basic research, which is not geared to such applications. Both basic and applied research are promoted within the framework of oriented research, since only a combination of the two forms of research can contribute to the solution of impending problems (Häberli 1995).

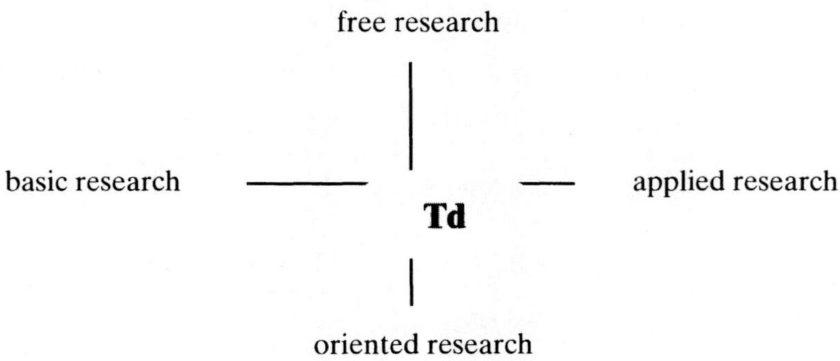

Figure 1
Types of Research (Td = Transdisciplinarity)

Transdisciplinary research is a form of action research. Participation and learning cycles have to start from beginning. The personal involvement of partners from practice and other audiences in the process of knowledge production is more effective than older models in which experts first produce knowledge then teach "the others." If involved groups are "taking ownership" of the outcomes of a new "common research," the chances of follow-up for acceptance and decision making are enhanced and new boundary conditions are created (*Gerber; Pokorny*).

Transdisciplinary research has both strengths and limits (Tab. 1). The complexity of problems of societal concern also confronts limited funding and human resources. The growing gap between demands on science and available resources means that priorities in publicly funded research are being renegotiated. Greater focus must be placed on fields of societal concerns, rendering solutions based on "blue sky" research increasingly illusory. Transdisciplinary research, together with practice, will continue to reveal new preconditions for successful innovations and their most likely emergence (*Böttcher; Jansen*).

At the same time, research at the fringes of natural- and engineering science and entire fields of social sciences is becoming so complex that feasibility becomes doubtful. Only mixed, transdisciplinary groups will be able to expand the limits of their research to a realistic model of real-world conditions, while reducing its complexity to insure an undertaking remains feasible (Hofer and Stalder 2000). Knowledge partners need to know in time which partners "from the other side" really want to know and what information will help them expand within their own fields. The

Table 1
Limits and strengths of transdisciplinarity for society and for science

Subsystem	Limits of Transdisciplinarity	Strengths of Transdisciplinarity
Society	1) time consuming processes	1) to identify problems early
	2) openness	2) to democratize science
	3) danger of being a study object	3) to produce social robust knowledge
Science	4) context-dependent results	4) to allow contextualization
	5) danger of being instrumentalized	5) to speed up implementation
	6) disciplinary socialization of scientists	6) to integrate different kinds of knowledge and produce new insights and knowledge

idea that first we have to understand "the whole universe," then derive answers to societal problems, also becomes more illusory.

As Olaf Kübler, President of the ETH, Zurich cautioned in the opening ceremony, as well as Charles Kleiber in Chapter 3.1, and Pieter Leroy and Kirsten Hollaender in Chapter 6.1, there is a danger of "transdisciplinarity" becoming a buzzword. What counts are outcomes, the products of transdisciplinary processes. It is clear that transdisciplinary cooperation needs time, even years, to develop, prompting Charles Kleiber to emphasize "the virtue of patience." In the evolving new contract among democracy, market economy and science, transdisciplinarity can play a decisive role in influencing relations in this *ménage à trois*. For that reason, it is not just a procedure but has a value as well. It functions as a messenger between civil society and market economy in the accelerated race of innovation in science and technology. In this historical change, science does not play its role without undergoing change itself.

Where is Transdisciplinarity Needed?

Transdisciplinarity is needed in many areas:

- in fields of great human interactions – such as agriculture, forestry, large technical interventions, industry, traffic, and megacities – with natural systems – such as water, soil, vegetation, atmosphere; and in the management of those natural resources (*Guichun et al.; Robledo and Sell; Vahtar*);
- in fields of major technical development such as nuclear- and bio-technology, genetics, and transport systems, where early participation for

clear views of possible consequences and public acceptance is essential (*Büchel; Dahinden et al.; Mirenowicz; Wackers and Schueler*);
- in the development context, where reflective action is crucial (*Baumgartner; Geslin and Salembier; Flury; Jenny and Baumann; and in Mutual Learning Session M18, Dialogue Session D07*).

In addition, transdisciplinarity has already proved effective in fields of great societal concern, including aging, banking, education, energy, health care, migration, nutrition, pollution, sustainable development, urban- and landscape development, and waste management. In these fields, social-, technical- and economic developments interact with elements of value and culture. Under these circumstances it is often difficult to find ways out of dead-end situations shaped by old habits and group interests. (For exemplars, see winning contributions of the Swiss Transdisciplinarity Award in Part 5 as well as *Aiking et al.; Alio et al.; Baumann; Baumann-Hölzle; Flüeler; Fässler; Herren and Mueller; Höpflinger et al.; Keitsch and Wigum; Kiwi-Minsker, et al.; Kyrtsis; Mogalle; Perrig-Chiello, Höpflinger and Perren; Toussaint et al.*).

Beyond these "big fields" of challenge, there are other domains where transdisciplinary or Mode 2 research (Gibbons et al. 1994) is fruitful, because it moves not just from a problem to a solution but also backwards from potential users of results to definition of a problem and design of a research project. This is particularly true in two instances:
- When the primary goal is to profit first from existing, but widespread and often controversial knowledge, and where computer-aided information will be increasingly useful (*Schweizer*);
- When existing general knowledge must be applied to very specific but inherently complex cases, as when localized environmental standards for soil reclamation have to be set or the decline of fish population in whole river systems has to be understood (*Holm; Ouboter*).

Developing Transdisciplinary Practice

Clear Goals

Transdisciplinary projects are promising when they have a clear goal from the beginning, a goal on which representatives of different disciplines agree. Any "hidden agenda" is an obstacle in the process of achieving constructive collaboration. A heterogeneous team of workers with different disciplinary and institutional backgrounds can reach its objectives only if it firmly pursues definite and common objectives and if every member of the team has a reasonable chance to get a fair share of commonly-produced results. Transdisciplinarity isn't without conflicts. Researchers and funding organizations will struggle about the optimum between "bottom up" and "top down" in management. Researchers and partners from practice and other audiences also have to agree on a common definition of the problem, to negotiate relevant themes and questions to be answered as well as methods to be followed and desirable outcomes (*Geslin and Salembier; Kasanen; Rubin de Celis et al.*).

Careful Preparation

Transdisciplinary projects require careful preparation. A balance must be achieved between widening limits to the complexity of a "real-world" problem and simplifying the research so that it is "feasible." This delicate task may even be worthy of a pilot study. Aspects of contents, philosophy, and organizational matters must be discussed, clarified, and focused on common concerns before starting. The more attention to these topics at an early stage, the smoother the ensuing collaboration and the better the ultimate results. Transdisciplinarity is a means for a goal, and not an end in itself. Whatever can be reached by easier monodisciplinary procedures should not be burdened with the complications of transdisciplinarity. Special attention must also be given to team-building and team processes. Only a genuine team, which is more than a coincidental gathering of specialists, will achieve the new insights a transdisciplinary process can nurture (*Defila et al.; Hoffmann-Riem; Jeffery et al.; Loibl; Van de Kerkhof and Hisschemöller*).

Competent Management

Competent management will minimize frictions among members of a team and foster the freedom needed for creativity. Management should organize the flux of mutual information and meetings in well-chosen intervals. Objects of common concern need to be focused on and team process stimulated through realization of common products. For successful transfer and implementation, the perspectives of both knowledge producers and those affected by the outcomes must be heeded. Users have their own agendas that have to be taken into account. Transdisciplinary research in big and complex teams is a delicate undertaking, requiring a permanent and responsible management to guide the common process and clear obstacles. A specially attributed "observer of the common learning process" or even an external and professional supervisor will help smooth the collaboration process and make it most efficient (*Gros; Hirsch and Jemma; Martinez and Schreier; Schübel*).

Sufficient Funding for Cooperative Work

Transdisciplinary projects cost more than simple disciplinary projects. Because funding agencies are often unfamiliar with these new forms of integrated research projects, they may have difficulties with the volume of the entire undertaking, especially costs for collaboration and management. Collaboration and management, however, is not the place to economize. Sufficient funding for preparation, management and collaboration is vital to creating the necessary conditions for genuinely transdisciplinary results. Producing new, synthesized and consensual insights as a result of a goal-oriented, long-term, broad and carefully-managed common process is the heart of a transdisciplinary undertaking. It cannot be done cheaply. Scarce funding for management and team process means the undertaking will be loaded with obligations and complications, without opportunities to produce the special outcomes expected from a common process. A collection of disciplinary contributions united under the same book cover is not a transdisciplinary achievement.

Table 2 provides a summary checklist of the main elements for success.

Open Questions and Future Research

The conference clearly showed there is great potential in transdisciplinarity and a growing community around this subject. Yet, it was equally evident that theory of transdisciplinarity and consensus on quality standards for transdisciplinary research are still under construction. Some of the most urgent questions are:

- How can crucial problems requiring a transdisciplinary mode be identified?
- How can conditions for fostering socially robust knowledge be developed?
- What is good practice in transdisciplinarity?

Table 2
Checklist for successful transdisciplinary research (in: Haberli and Grossenbacher-Mansuy, 1998)

1. Goals	❑ Clear, compatible goals
2. Preparation	❑ Joint formulation of research questions ❑ Problem analysis ❑ Method selection ❑ Product planning ❑ Scheduling and budget planning ❑ Team formation and networking ❑ Forms of cooperation and conflict regulation
3. Management	❑ Support for internal and external information flows ❑ Regular meetings ❑ Constant overview – taking stock and thinking ahead ❑ Targeting of joint products ❑ Knowledge implementation ❑ Responsible research leadership ❑ Learning-process monitors
4. Resources	❑ Necessary, suitable transdisciplinarity ❑ Proper comparison ❑ Adequate funding of group process ❑ Co-funding ❑ Advanced funding of preparatory work
5. Environment	*Research* ❑ Cooperative behavior and implementational competence *Science* ❑ Dialogue with practitioners ❑ Openness to transdisciplinary research ❑ Advisory agencies and platforms ❑ Career incentives ❑ Publication channels and organs *Practice* ❑ Expectation of science ❑ Understanding for academic research *Funding* ❑ Creation of transdisciplinary promotional instruments ❑ Development of new evaluative criteria *Methodology* ❑ Epistemological research on transdisciplinary processes

- How can knowledge integration be achieved successfully?
- When should practice step into a transdisciplinary research process?
- Is transdisciplinary research a team process or is it possible to do it alone?

- What would a clarified theoretical concept of transdisciplinarity look like?
- What would a review system and quality control for transdisciplinary projects look like?

Promoting Transdisciplinary Research

Openness on the Side of Academia

Joint problem solving among science, technology and society means that universities and other publicly-funded research institutions need to open themselves to collaboration with practice and with groups from outside academia. They must invest their own financial and personal means to create and to maintain this collaboration. They must introduce training in transdisciplinary research as an important component in the educational preparation of their students. They must create secondary structures in which members of their teaching and research staffs, as well as students, can temporarily join and work in transdisciplinary projects (see *Scholz and Marks, Chapter 6.2*). Universities should create chairs for teaching and developing transdisciplinary research methodology. They should also create and apply new and adequate evaluation systems, taking performance in transdisciplinary research into account in career evaluation as they now do with specialized disciplinary research. They also need to recognize the performance of persons from outside academia, who are distinguished in promoting collaboration between academic and non- academic research in academic prizes, titles, and medals.

Openness on the Side of Practice

Collaboration with publicly-funded research is not always evident on the side of practice, either. Practice is interested in ready solutions and short-range applicable results. Universities have longer time scales and preferences for basic research and general analysis. Collaboration between practice and groups from other parts of the society with academia will only develop if the side of practice works towards a new understanding of academia. This change will entail accepting consultation and proposals from academia, agreeing to be a partner in research projects, being involved in experiments, acting as a teacher or a tutor, giving time for research, waiting in fundamental decisions until results of research are available, and making efforts to understand fundamental questions of research. Both partners can learn from each other: Collaboration in a transdisciplinary research project requires partners from practice to open their horizons, producing new ideas for getting out of dead-end situations and even developing new products to position themselves on the market. For participants from science, collaboration results in new views and ideas, better understanding of the "real world" and testing and adapting of their theories, and contributing to products society needs. Practice, though, has to demand achievements from research. In the long run, both sides have to have their profits and collaboration must pay (*Markard and Wüstenhagen*).

Platforms for Contact and Promotion

Joint organizations among universities, other publicly-funded organizations and non-academic organizations create platforms and advice centers for transdisciplinary research. To start with, those centers can be created by universities. The task of such a "facilitating organization" is to help in making contacts, starting projects, creating efficient teams and management, and getting funded. In these new organizations, representatives from economics, politics, administration, the general public and research can meet and draft projects. Such organizations must be able to promote transdisciplinary projects efficiently in the critical starting and finalizing periods, with easily obtainable seed and finalizing money. To insure effective staff organization, it is important to have an extensive social network and generous organizational and financial autonomy. Developing methodology, teaching, and disseminating best practices is another important activity. Platforms for contact and promotion can be connected with a chair for development of transdisciplinary research methodology. Similar platforms have been established successfully for special fields of collaboration and promotion of new ways of dealing with urban or landscape development, in tourism, and in soil quality management (*Habisch and Meister; Hirsch Jemma and Stalder; Schenkel; Van Veen and Ouboter*).

Adequate Forms of Evaluation

Transdisciplinary projects should be evaluated in a different mode than disciplinary projects. Transdisciplinary tasks are generally bound to a certain context: in place, time, socio-economic and political surroundings, and main partners and actors. Evaluation needs to take into consideration this special context. It cannot be limited to general, global criteria. Beside criteria of general methodology and disciplinary research, which must be "state of the art," criteria such as special context of application, team process and participation, outcome and problem solving should be taken into account. Evaluation also, especially in goal-oriented transdisciplinary research, has to serve a purpose. Evaluation of a transdisciplinary project differs, though, by stages: whether a project is preliminary to launching, in the launching phase, in the middle of its course, in completion, or after completion. Moreover, if the goal of the evaluation is the impact of results in "real-world" activity, there is often an important time gap between demand for an application in "real-world" practice and availability of scientific results. Delays of years and even a decade are not unusual and cannot be blamed on the transdisciplinary project (Defila and Di Giulio 1999; *Defila and Di Giulio; Krott; Rege Colet; Spaapen and Warmelink*).

Creating Favorable Institutional Structures for Transdisciplinarity

Transdisciplinarity as a Part of University Education and Training

The future will be more dynamic. The time when young people acquire professional training up to the age of about 20 or 25, then stay in this activity to the age of retirement, has passed. Therefore, training in how to use many forms of information and knowledge, how to work in teams with experts from many different fields, and how to learn and adapt quickly and permanently to ever-changing situations are new imperatives for preparing future professionals. The Internet, distance learning and other generally available sources of knowledge will accelerate this development.

Universities and other higher education institutions must adapt, as well as other institutions of the educational system. In his keynote address, *Uwe Schneidewind* (*Chapter 3.5*) elaborated on the organizational deficits in today's universities and how they can be filled. Besides basic preparation in a specialized field, knowledge acquisition, selection and management, and collaboration in changing and time-limited teams, will be growing at all levels, including higher education. This new education requires more than just adding new lectures or training exercises in basically unchanged disciplinary courses. It means creating and actively maintaining a secondary structure where transdisciplinarity is practiced, developed and trained as an important part of students' basic education.

Furthermore, this secondary structure has to be an important part of university life. [Rudolf Häberli argued it should be as much as 30 % of total academic budgets and activities.] It cannot be limited to a marginal existence of small percentages. The permanent staff of the secondary – transdisciplinary – structure should be small, composed of initiators, facilitators, organizers of platforms, coaches, theorists and teachers of transdisciplinarity. Most people working in this secondary structure will be there for a limited time, leaving their disciplinary structure for one to several years then going back to the original structure after the end of the transdisciplinary project. (See *Scholz and Marks, Chapter 6.2*, on creating Transdisciplinarity Laboratories, as well as *Abbasi and Gnanam; Abou-Khaled et al.; Canali; Cslovjecsek; Frischknecht et al.; Gökalp; Grütter et al.; Keitsch and Wigum; Kuebler and Catani; Künzli; Lima and Rutkowski; Schulz and Baitsch; Villa et al.; Yonkeu et al.).*

Adequate Funding Structures

Realizing this new and time-limited transdisciplinary research will require significant financial resources. Most of this money, though, should not be "new money," rather a redirection of funds. The goal is to make researchers work in a different way, in transdisciplinary teams and projects. The money may be taken from traditional disciplinary structures and placed in a new "transdisciplinary pot." From this pot, members of traditional disciplinary structures can regain money by participating in transdisciplinary projects. The new allotment will support platforms and "facilitating organizations" for contact, promotion and education in transdisciplinary research.

In addition, special divisions of funding agencies should be created with the exclusive task of funding transdisciplinary research. One of their charges would be to launch large "pioneer" programs in main fields of interest to society, with the goal of promoting transdisciplinary, interdisciplinary, and interinstitutional arrangements, including practice and audiences. The SPPE is an exemplary model. These divisions can also support the activities of platforms and "facilitating organizations" of universities and of mixed academic and non-academic organizations. By concentrating on funding transdisciplinary activities, the divisions will develop finally, and most importantly, new and adapted systems of calling for research proposals, evaluation, control and dissemination, and rewards and incentives. Large pioneer programs sustained by special funding divisions will be major factors in creating a new culture of transdisciplinary research.

Transdisciplinary research, of course, should not be done only when external money is available. To reiterate, universities and other publicly-funded institutions should contribute significant percentages of their own budgets. Additionally, partners from outside academia, from industry, business, administration, NGO's and other parts of the society should make financial contributions. Through experience, they have learned that transdisciplinary collaboration is fruitful and necessary. Joint research and joint problem solving will be based on joint funding, leading in turn to joint responsibility and real "mutual learning."

If the Incentives are Right, Everything Goes Right

More and more, forward-looking leaders are engaging in transdisciplinarity. Yet, it is still an outsider in organizational structures. Therefore, special support is still needed and structures still need to be changed. Transdisciplinarity remains new, therefore weak and vulnerable. It is confronted with the challenge of passing from a niche market to a mass market. The transition from one established system to another is difficult. Transdisciplinarity needs opportunities to grow and to fortify, to learn by doing, to progress by learning from failures, and to progress all the way. The most important element is recognition of this form of joint research and learning by society. Influential individuals need to share common values and have regular discussions to nourish those values. Platforms and "facilitating organizations" and action- or strategy centers must also actively promote and develop transdisciplinary research. With experience and success, transdisciplinarity will become more widely known. As a result, more funding can be procured and achievements by transdisciplinary teams recognized. They must be recognized not only with kind words, but also with diplomas, rewards, titles, financial success, and professional careers. Once transdisciplinarity is recognized as a normal and useful activity of universities, large research institutions and other parts of society and once it is solidly installed in the main funding organizations, special support will not be needed. The whole system will regulate itself. At present, though, there is still some way to go (*Hvid and Homann Jespersen; Lawrence*).

Conclusions: Transdisciplinarity: Joint Problem-Solving among Science, Technology and Society

With scarcity in public money, expenditures for education and research are constrained. Efficient investment of public money is called for. Too often, however, austerity measures lead to concentration on traditional fields and methods. Retiring to the past is not the right answer for the challenges of the future. As in other sectors of politics, orientation to the "customer" is important in research. Transdisciplinary research has to serve society. To repeat, it is not a goal in itself (*Perrig-Chiello; Perren et al.; Sundin*). The central question is: How can research be organized to solve real-world, practical problems as effectively and cheaply as possible. Transdisciplinary joint problem solving among science, technology and society is an answer to the demand for greater customer, stakeholder, and user orientation of research and for raising its level of utility. "Free" or "basic" disciplinary research and science will retain its value. Transdisciplinarity as a new form, though, will grow at its side, becoming more and more important.

Taken together, the contributions to the conference suggest five major lessons.

- First, a truly good practice for transdisciplinarity already exists.
- Second, the Mutual Learning Sessions, especially, demonstrated that there are various and promising fields for transdisciplinarity.
- Third, transdisciplinary processes have a potential to crack locked-in situations.
- Fourth, questions still remain, especially about theory (epistemology, methodology, and organization) and about common quality standards.
- Fifth, learning, promoting and adapting structures are vital activities.

Further progress is needed, but also expected, in each of these directions.

Establishing new modes of proceeding and changing structures and incentives are crucial to the future. The conference introduced the Swiss Transdisciplinarity Award as an incentive to recognize pioneers and successful researchers in this arena. (See *Part 5*) The Swiss Academic Society of Environmental Research and Ecology has been active for years in this field and, as an immediate follow-up to the conference, has created a new international network of transdisciplinary researchers (*Förster and Hirsch Hadorn*). "Science et Cité," a newly-created foundation aimed at improving relations between research and society in Switzerland, is also promoting transdisciplinarity. Its President and founding member, State

Secretary Charles Kleiber, has proposed a transdisciplinarity collegium to promote the idea of transdisciplinarity. Alain Bill, from ABB, offered an additional proposal. The International Transdisciplinarity Conference may be the starting point for a biannual conference series. Last but not least, this book is an important step in building a platform for follow up, bringing conference participants together again around the task of achieving a common goal. Managing complexity and joint problem solving among science, technology and society is crucial to meeting the challenges of the 21st century. We need to continue working towards this common goal and improving the means of achieving it.

SAGUFNET – the SAGUF network for transdisciplinary research

Transdisciplinary research is a promising way to tackle complex problems that are beyond the scope of disciplinary research – e.g. environmental or societal problems. Today, transdisciplinary research is moving towards professionalism and reinforcement of the research community is imperative. It is the latter, general goal that sagufnet will pursue by the following means:

- Organizing 1–2 workshops per year on topics that concern transdisciplinary research
- Publishing results of the workshops and distributing them among the networkers
- Supporting information exchange among networkers
- Integrating other networking activities in and out of Switzerland

Who should participate? SAGUFNET will be of interest to all researchers in transdisciplinary projects, especially young scientists. Participation of extra-scientific stakeholders top transdisciplinary research is also encouraged.
To become a member simply subscribe. Membership is free of charge, a minimum fee to cover expenses will be raised for special events.
Sagufnet is established by the Swiss Academic Society for Environmental Research and Ecology (SAGUF) and managed by the MGU Coordination Center at the University of Basel. SAGUF is a mem-

ber of the Swiss Academy of Sciences (SAS) and an associated member of the Swiss Academy of Humanities and Social Sciences (SAGW). MGU promotes inter- and transdisciplinary education and research in different areas since 1992 and has established the corresponding programs at the University of Basel.

Write to
SAGUFNET
c/o MGU, Universität Basel
Postfach
CH-4002 Basel
www.unibas.ch/mgu/sagufnet
E-mail: sagufnet@ubaclu.unibas.ch

SAGUF
c/o Dr. Christian Pohl
ETH Zentrum, HAD
CH-8092 Zürich
www.saguf.unibe.ch
E-mail: saguf@umnw.ethz.ch

2

Introduction

2.1 Why a Globalized World Needs Transdisciplinarity

by Alain Bill; Sybille Oetliker, CASH, Zurich; Julie Thompson Klein

Abstract
Members of the opening panel, the chief editor and the moderator reflect on why transdisciplinarity is needed, obstacles to its realization, and critical factors for success. They represent a variety of sectors, encompassing politics, industry, wildlife and natural resource management, and technological innovation.

Transdisciplinarity is necessitated by new types of problems. These problems arise from economic and technical globalization; their social, political and cultural impacts; and interrelated global and regional changes in the environment. Because they are complex, the problems can only be solved through the cooperation of many sectors of society and, in today's globalized world, with an intercultural attitude. The panelists and members of the audience, moderator Sybille Oetliker reflected, have different expectations. Yet, they also have common questions. What kind of problems should or can be solved by the transdisciplinary approach? What specific solutions can be found? Why, in reality, does transdisciplinarity not work often?

Demands of Society on Science

Moderator:
Sybille Oetliker, CASH, Zurich

Panelists:

- Martine Brunschwig Graf, Directeur de l'Education, Geneva
- Tony Kaiser, Head of ALSTOM Power Technology Ltd, Baden-Dättwil
- Claude Martin, Director General WWF, Geneva
- Thomas von Waldkirch, Director Technopark, Zurich

What Problems Require Transdisciplinarity, and How Can They Be Solved?

Martine Brunschwig Graf, as a representative of politics, must find solutions to societal problems. She is a state counselor in the republic and canton of Geneva, in charge of the Ministry of Education and the Ministry of Military Affairs. She was educated as an economist and worked previously for a Society for Development of the Swiss Economy.

A Problem: Offering medical help to all people living in Geneva. How can we bring together medical and social knowledge? How can we reach a situation that is accepted by a large majority? How should we train future doctors to act in the best way?

A Solution: All types of experts have to work together: scientists and doctors, who can evaluate which medical services are necessary; social workers, who know how to bring services to the people; and economists, who must find financially-feasible solutions.

Comment: In political affairs, a realization is growing today. Problems cannot be treated without realizing inter- and transdisciplinary approaches. There are other examples. In the area of drug policy, security, health prevention and care, social measures must all be taken into consideration. In family policy, taxation and education, social considerations arise. In neighborhood development, the quality of "environment" is a complex, not a narrow, concept.

Tony Kaiser, as a representative of a multinational enterprise, must compete in a global marketplace by creating new products that are quicker, cheaper and better. He is Director of the ALSTOM Power Technology Center in Baden-Dättwil and Director of a Technology Center in Heidelberg, Germany. Kaiser was educated in the field of physical chemistry and, earlier in his career, worked on polymers in the development laboratory of Brown Boveri Company.

A Problem: Finding environmentally-friendly high-voltage switch gears to replace mechanisms that emit dangerous greenhouse gases. How can the complex interplay of factors be managed?

A Solution: All factors must be weighed using an approach called Life Cycle Analysis. All phases of research must be considered and cooperation occur across science, engineering, modeling, manufacturing, project management, financing and even follow-up services and support functions. The need for a broad range of specialists underscores the importance of greater cooperation with universities.

Comment: This approach was also used to determine the most economically-efficient and environmentally-clean way of generating electricity from fossil fuels. Life Cycle Analysis provided a comparative, holistic picture of the feasibility for gas, coal, or oil options. Gas emerged as the best choice that met both economic and environmental criteria.

***Claude Martin*, as a representative of an internationally active NGO, is involved in conservation efforts across the entire world. Since the 1970s, he has worked in different ecological projects in India and Ghana. The World Wildlife Fund (WWF) is currently in about 100 countries, employing approximately 3,000 specialists in not only biology but also social sciences and resource economics. WWF works in local field-orientations as well as on global policy and educational programs.**

A Problem: Stopping deforestation in Indonesia. How can over-capacity be replaced by sustainable levels of forestry? How can the complex set of economic, political and social factors be dealt with?

A Solution: There is no "silver bullet" solution but, given that the problem is rooted deeply in social-political contexts, a serious national and provincial stakeholder process is required. WWF operates as a "broker" in alliance with World Bank issues. Many experts must be brought together, not only biologists and environmentalists, who understand the detrimental effects of deforestation but also social scientists.

Comment: Three other problems call for the same approach – climate change, management of marine fisheries, and the looming freshwater crisis. International organizations such as WWF function as an "early warning system." Yet, they are often isolated or dismissed. So, patience is needed.

On the positive side, the problem is recognized, but representatives of political authorities must also be convinced of a solution and international organizations involved.

> *Thomas von Waldkirch*, as a representative of an institution for promoting technological innovations, helps to create new enterprises and products for the marketplace. In his role as Director of the Technopark, he facilitates technology transfer by bridging gaps between different technological disciplines and entrepreneurial aspects. Von Waldkirch was educated in the field of experimental solid state physics and worked previously at the ETH.

A Problem: How to bring together the know-how of different specialties of technology, economics, human resources, design, public relations and other needed areas.

A Solution: No single human being possesses all the knowledge required. The Technopark provides a permanent platform for "networking competencies" across innovation, transfer, and production. Partners find the most effective ways of fitting together. Many important innovations, such as Magnetic Resolution Imaging (MRI) technology, have been found by making novel combinations of expertise. The Technopark has fostered collaboration between Sulzer Electronics and ETH, Zurich in developing and introducing a linear-motor drive in machine construction (automatic machinery and robots); a specialized wine-trading firm via the Web <www.vinoplus.ch>, resulting in an attractive website, high-quality products and reliable logistics; and a small Doppler instrument for vascular diagnosis and childbirth, as well as an Internet-Platform for networking new products and creating new team-based enterprises.

Comment: We must move from an "information society" to a "cooperative society." This will require, to recall Olaf Kübler's address, curiosity, as well as the ability to interface with other disciplines, and a willingness to cooperate and to find partners who fit together. All aspects must be considered, including economics and human resource management.

What are the Obstacles to Transdisciplinarity?

The dominant mode of working in science has been within isolated disciplinary frameworks, using specialized theoretical concepts and analytical tools. Moreover, social science approaches have not been incorporated seriously into the mainstream of environmental research, and vice versa. Environmental considerations are excluded from the mainstream of social sciences, where a deep gap is also cutting across individual disciplines. Why, despite powerful and pressing arguments for transdisciplinarity, do disciplinary modes still dominate?

Tony Kaiser: There are many limits in the industrial environment – not enough time, money, and resources, plus a competitive marketplace that restricts sharing of information. Companies are also profit driven. An interdisciplinary approach and environmental focus work best when ecological and economic drivers are aligned. Questions of external costs, politics, and legislation are not driven by industry alone. Sometimes a particular situation propels a change. In the case of legislation on greenhouse-gas emissions, limits were given a strong "value" dynamic and market forces responded.

A Sample Case Study in Solving Complex Problems: ABB Alstom Power

Complex processes resulting from a variety of needs and functions are driving society. Complex systems thinking is at a critical stage, emerging from the promise of rhetoric and theory into arenas of implementation ranging from organizational structures and management paradigms to engineering design and logistics. ABB distinguishes four basic dimensions (Fig. 1):

- scientific processes, referring to levels and modes of research and innovation
- economic processes, referring to levels and modes of production and consumption of goods
- social processes, referring to individual evolution within an environment
- political processes, referring to issues such as participation in, and responsiveness of decision-making processes, plus the capability of institutional arrangements to implement policy strategies and to accommodate to changing conditions.

Successful innovation cases in environmental and process industries are good models of using entrepreneurial and other relational

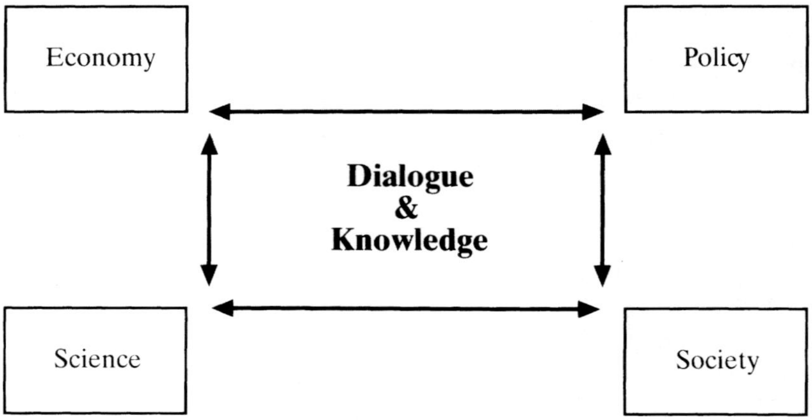

Figure 1
View of the Different Processes

dimensions. In most instances, know-how has been developed with contributions from many institutions (e.g. research institutes, enterprises, state authorities) and their professionals. Research and innovation are becoming more typically transdisciplinary processes.

In Figure 2, the solution path for a transdisciplinary team is highly interactive and back-coupled. The first challenge is definition of the theme (e.g. the problem that requires solution). The second is the process whereby the problem is analyzed and a solution developed, leading to a result. The strategy for structuring includes the leading vision as well as the means and capacities of actual or potential participants. On this basis, an action plan is defined, setting priorities, resources, target dates, etc. Follow-up insures feedback on the idea and the solution that is being generated. At the research stage pluridisciplinarity proves to be important. At the development and production stages, concrete possibilities (e.g. a pilot plant, etc.) and a sustained commitment to realization are essential. At the sales and service stages, the real demand and a favorable legislative environment will be decisive. The team coach has a crucial task: leading the team through the entire operation in a manner that creates a learning organization. This environment is characterized by members who know how to search, ask questions about the theme, reflect on the

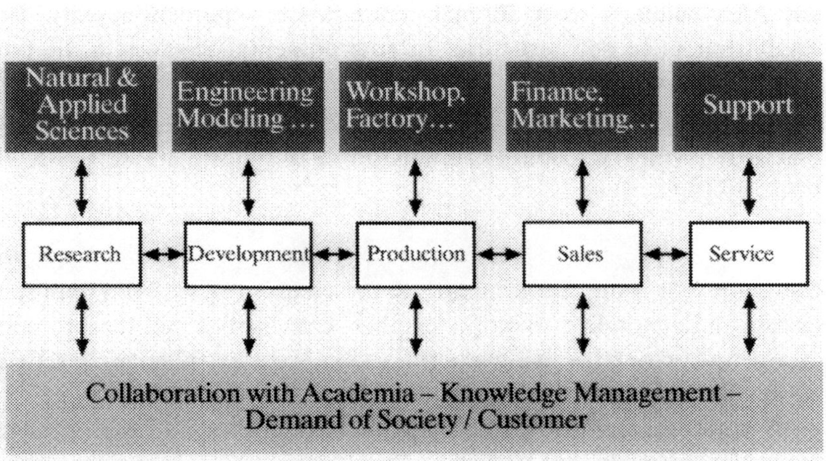

Figure 2
Current Complex Problem Solving Path by Industry

team's strengths and weaknesses, and adjust goals of the common undertaking throughout the project life cycle. The project manager must be an individual who is acknowledged by all team members. Furthermore, the quality of relational dimensions is of paramount importance. In keeping with transdisciplinarity, mutual comprehension and appreciation among scientists, technicians and potentially users have proved to be major success factors.

Thomas von Waldkirch: I propose two important terms. The first is "impedance gap." In transmission technology, it refers to transmitting from one place to another. If the impedance is not the same on both sides, a signal does not go over. The parallel in reality is lack of a common language. As the basic sciences become more complex, more time is needed in universities for disciplinary education and research. At the same time, industry and the economy create pressure for reducing that time. The second term is "distrust," specifically distrust of the scientific quality of transdisciplinary research. It risks being eliminated by established criteria of evaluation. However, there are many brilliant examples of discoveries and breakthroughs that would not have occurred if traditional criteria alone were imposed, such as the Nobel Prize recipient Alex Müller's work in high-temperature superconductivity. The establishment of new structures of environmental sciences in the university is another example. This challenge became more complex with the increasing politicization of areas such as alternative energy, traffic and gene technology. Politics should not be in the research process but keep out of it.

Martine Brunschwig Graf: Intervention in science, indeed, is not the politician's job. Nor is the politician able to do science. Yet, we need both specialists and generalists to work together. Universities still tend to educate people in one speciality at a time when transdisciplinary capacity is all the more needed.

A minister responsible for education is keenly aware of the need to move toward inter- and transdisciplinary preparation across the entire educational system. It is not possible to be good in everything. But, everyone can learn to work together through networking. Part of the meaning of "global" thinking is working across not only countries but also disciplinary specialties, compartments of professional training, and official classifications in organizations. The situation today is changing in professional and academic education. Inclusion of a problem-based approach in medical training is a noteworthy example. More must be done though middle- and long-term thinking is also required, not just short-term thinking.

Claude Martin: Lack of preparation for transdisciplinary research is a major problem. Analyzing a problem and arriving at a solution are not so challenging as managing the solution and implementing it. Martin's own Ph.D. experience returned to mind. Working in India, as a young biologist during his first field research, he grappled with the threatened disappearance of a particular deer species. Analysis revealed a complex set of reasons and circumstances, including human degradation of a particular substrate of vegetation the deer ate. A "generalist" species of deer was then competing with the endangered "specialist" species. (In biology as opposed to academic, Martin quipped, the generalist wins over the specialist). The problem could be corrected with counter-active management strategies. Yet, upon returning to his university in Zurich, a bigger challenge loomed. The zoologist and the botanist on the Ph.D. committee, lacking knowledge of each other's specialities, were dismissive. The student, like all transdisciplinary problem solvers, wound up having to engage in conflict reconciliation.

Concluding Perspectives of Audience and Panelists

Felix Rauschmeyer, an economist working in Leipzig, Germany, asked whether Socioparks, along the lines of the Technopark in Zurich, might be created for the WWF and others – in order to promote collaboration among social scientists and other kinds of specialists.

The goal, *Martine Brunschwig Graf* urged, should not be to have just one park but getting a center for "perpetual discussion." People from politics and from universities should not think in terms of a war but a network. *Thomas von Waldkirch* emphasized the need for people to trust each other. They cannot do so in cyberspace alone. They must see each other in person and get to know each other in order to build trust over time. One model for working together, not separately, is the spatial merging of the federal technical institute and the university in Lausanne.

Sunita Kapila, a program officer with the International Development Research Center's office in Nairobi, noted that the problem in Indonesia of illegal tree cutting is symptomatic of a number of problems in the South, due to sacrificing natural resources to the market principle. Since the World Bank, like the International Monetary Fund, has perpetuated the market economy, what is their role? Relatedly, since people do not want to give up their economic livelihoods, what can be done? Also, how can stakeholder participation truly occur in non-democratic and corrupt government? It is not possible to be "pristine." What, then, can be done?

The World Bank, *Claude Martin* replied, has changed under new leadership that instituted a different forest policy. There is new networking across institutions. In Indonesia, even illegal timber cutting generates jobs, but only in the short term. It is not possible to be absolutely "clean," even in a nation such as Switzerland. So, cooperation in finding solutions is required. In working together, all partners must make compromises. The stakeholder process has also been implemented in dictatorships and corrupt countries, such as Cameroon, when the WWF accepted an important principle. Sometimes, you need to accept unfamiliar ways of operating.

Martina Keitsch, from the Norwegian University of Science and Technology, asked how both quantitative and qualitative approaches can be used. In her field, ethics, there is a method for working with the complex issue of responsibility. Yet, even when working together in ethics, there is a "quality loss." How can disciplines be made to change and to understand each other?

A previous question from the audience had prompted several reflections on the importance of responsibility and ethics. *Tony Kaiser* emphasized the importance of each individual assuming responsibility. Everyone must be accountable and reliable. The vague exhortation to behave ethically reminded *Claude Martin*, responding directly to Keitsch, of the common coffee area in workplaces. Everyone agrees in principle to clean up, but it is a mess anyway. If you identify a problem clearly then place it before a wider audience, engaging a "public" truly through a stakeholder process, people can't hide in an ivory tower. A target-driven approach leads to a better transdisciplinary approach rather working than sector by sector.

Parting Thoughts

Thomas von Waldkirch: Universities have to change their policies and develop transdisciplinarity, without imposing limits. First, we must cross boundaries. The ETH managed projects of a certain size (Polyprojekte) are good models of collaboration. Yet, they need to be managed well, so that new quality standards for transdisciplinary work are defined.

Claude Martin: Management of complex problems is impossible without a public debate of some type. Only then do you get accountability across borders, but scientists are shy of public dialogue. The challenge for institutions such as ETH is to identify problems, formulate clear aims, and public dialogue on the possibilities of solutions. Then transdisciplinarity will be opened up and fostered. We also need scientific managers who are capable communicators.

Tony Kaiser: We have gone deeper and deeper with our specialization. It is time to go broader and broader as well. We must learn to manage information and learn how to draw the right conclusions from data that is not yet usable information.

Martine Brunschwig-Graf: Finally, science is not something apart. It is in the society. We have to accept complexity. Truth can be everywhere. We have to look for it together. Each of us has to be very humble.

2.2 The Discourse of Transdisciplinarity: An Expanding Global Field

by Julie Thompson Klein

Abstract

The International Transdisciplinarity Conference in Zurich was a defining event in the evolution of a new discourse of transdisciplinary problem solving. The new discourse is linked historically with a new climate for participatory research on sustainability problems, changes in the nature of disciplinary research, and the emergence of new knowledge fields. The state-of-the-art represented by the conference is a truly global platform for future research and collaborative problem solving.

The International Transdisciplinarity Conference was a pivotal event in the evolution of a new discourse. Many of the keywords are not new, including "transdisciplinarity" itself (Fig. 1). Its origin is often traced to the first international conference on interdisciplinarity, co-sponsored by the OECD in 1970. The typology of "multi-," "pluri-," "inter-," and "transdisciplinarity" that emerged at the meeting established a new vocabulary of disciplinary interactions. Initially, transdisciplinarity was defined as a common set of axioms for a set of disciplines (Interdiscipli-

- problem- and solution-oriented
- complexity, heterogeneity
- participation of stakeholders, users, and decision makers
- collaboration, cooperation, partnership
- negotiation, mutual learning, joint problem solving
- cross-disciplinarity, boundary crossing, integrative, holistic

Figure 1
Transdisciplinary Keywords of Problem Solving

narity 1972, 25–26). The Zurich conference centered on a definition that has gained currency in recent years, especially in Europe.

A New Climate for Research

The new meaning was evident during the late 1980s in the movement, in Swiss and German contexts, from an older rhetoric of "going beyond disciplinary knowledge" to problem- and solution-oriented research coupled with participatory approaches (*Pohl*). Two keywords – sustainability and participation – are central to the new definition.

Sustainability and Participation

In 1970, UNESCO's program on "Man and the Biosphere" highlighted the impact of human activities on the environment. Growing attention to environmental problems led to the formation of new research and educational programs. The Department of Environment, Technology and Society Studies at Roskilde University, for example, has twenty years of experience with problem-oriented interdisciplinary teaching and research (*Hvid and Jespersen*). The 1987 Brundtland report of the World Commission on Environment and Development took a further step, defining principles for societal and environmental sustainability.

The year 1992 was a watershed. The UN Earth Summit in Rio de Janeiro established a platform for implementing principles of the Brundtland report. AGENDA 21, the final document of the Rio Summit, outlined sustainability as a future development strategy. Subsequent documents, such as the 1997 Kyoto Protocol fixing limits on green-house gas emissions, extended efforts (*Fussler, Ehmayer, Naveh, Rhon*). Sustainability became a global "catchword" for a holistic approach to issues of the environment and development (*Rais*). Responding to growing global concern, individual countries launched programs of research and problem solving. Sustainability also appeared on the agendas of particular sectors, such as banking, agriculture, and health.

As concern for sustainability expanded, an added change occurred –a shift from earlier emphasis on economic efficiency to inclusion of social justice and political participation (Becker et al. 1997; *Chapter 5.2.1*; *Becker*). Participation of stakeholders is not new. In Danish agriculture, for instance, the tradition of self-organized and cooperative development dates from the 19th century. In the late 1980s and the 1990s, new ideas for improving "planned participation" in environmental regulation emerged in Denmark and The Netherlands. In 1987, the Danish Board of Technology developed "consensus conferences," bringing public debate into technology assessment. Similar efforts followed in other countries, such as the Swiss PubliForums (*Nielsen, Agger, and Heinberg; Bütschi*).

The "participatory turn" in technology assessment (TA) is a major episode in this history. In the 1950s through 70s, TA was promoted by scientists and policy analysts as a rational-scientific tool for governmental policy and decision-making on science and technology. The primary approaches were scientific evaluation and cost-benefit analysis. TA was institutionalized by the US Congress in the early 1970s and at parliamentary levels in several European countries. The European model of TA always grappled with the challenge of integrating interests and values into assessment. During the 1980s and 90s, participation moved center stage and methods such as consensus conferences and scenario workshop models became part of institutionalized TA (*Chapter 5.3.6 by Nentwich and Bütschi; Joss*).

The new discourse marks the convergence of sustainability research and participatory methods with a new commitment to multi-sectoral collaboration. The European Union's Fifth Framework Program (1998–2002) is a benchmark of this convergence. Since World War II, a number of industrialized nations have targeted specific areas for problem-focused research, initially defense and space then, since the 1970s, areas of intense international economic competition such as engineering and manufacturing, computers, biotechnology, and medicine (Klein 1996, 173–208). At the same time other industrialized nations were targeting particular areas, the Fifth Program designated major social and economic issues, including employment and industrial competitiveness, quality of life, and problems related to public health, environment, and transport.

One particular effort highlights the added step that transdisciplinarity represents. The Fifth Program's project on "Quality of Life and Management of Living Resources" moves beyond earlier cooperation between academic and industrial researchers and their users to encourage involvement of other social actors, such as consumers, NGOs, health authorities, and patients' associations (Gerold 1.3.00). In the process, the definition of knowledge has extended beyond traditional canons of academic expertise to include "layman's" or "alternative" knowledge. This expanded form of knowledge is not pre-determined but emerges in the negotiation of multiple perspectives on a production, producing what *Gibbons and Nowotny*, in *Chapter 3.3*, call "socially robust knowledge."

New Knowledges

Transdisciplinarity is further implicated in the changing nature of knowledge. The list of contributors to the Workbooks (Appendix D) includes representatives of traditional disciplines, such as physics, chemistry, sociology, psychology, economics, business, engineering, and law. Historical separations of disciplinary cultures are still inherent in the way that universities function, but they are eroding and even becoming obsolete in some areas of research (*Weiss*). Gone are the days, *Colwell and Eisenstein* demonstrate in *Chapter 3.2*, when disciplines "go it alone." Evidence of crossfertilization abounds at all levels, from the gene to the gravitational forces of the universe.

Biology, to cite a leading example, became a boundary-crossing discipline due, in significant part, to the discovery of DNA in the 1970s. A veritable "cognitive revolution," this discovery has refigured traditional demarcations of physics, chemistry, and biology. It also fostered new fields of application (*Haribabu*). The geosciences are another example. At the empirical and methodological levels, new discoveries, tools, and approaches have changed the way research is conducted on a daily basis. At the conceptual level, the theory of plate tectonics fostered new linkages among the earth sciences.

The new discourse of transdisciplinarity draws on interdisciplinary developments in the disciplines while taking boundary-crossing a step further. In certain fields of geosciences, strategic analysis is called for, capable of identifying "real-world," user-oriented problems and demands (*Schönlaub*). The fields of development biology and developmental genetics raise ontological and moral questions about the status of "biology" in society, not just in the laboratory (*Neumann-Held and Rehmann-Sutter*). The discovery of DNA also has implications for other fields. DNA evidence has entered the courtroom. Lack of communication between forensic and judicial worlds, and misinterpretations of the value of evidence in court, require new approaches to training students for the law and forensic science (*Taroni, Nangin, and Bar*).

The list of contributors reveals an added development. Many participants are from new knowledge fields that only emerged over the latter half of the 20th century (Fig. 2).

A significant number of participants are in environmental and ecological sciences. Two "synthetic" disciplines are linked with a new paradigm in environmental research and education. Functional ecology evolved in the 1940s from synergetic interactions of biology, chemistry, physics, and earth sciences. In the 1960s, it discovered a new social role.

- ecology, environmental sciences, resource management
- industrial ecology, medical ecology, human ecology, social ecology
- life sciences, public health, cancer research, biotechnology
- sociology of knowledge, discourse studies
- science technology and society studies, future studies, regional studies
- cultural studies, media studies, communication studies
- information sciences, cybernetics, computer sciences
- systems sciences, knowledge management

Figure 2
New Knowledge Fields

Environmental education emerged from synergetic interactions of two kinds of contributions: (1) scientific knowledge, human sciences and environmental ethics; and (2) methodological contributions of education sciences and environmental psychology (*Achille and Antonio*).

As the list suggests, not all fields are the same. Some are practice-oriented, others more sociological and critical in perspective. Representatives of conflict studies, public and international health, resource management, regional studies, and cultural studies were among the participants. Many individuals came from new subfields that have strong sociological and critical interests, such as human geography and sociology of science. Generally speaking, the role of critique has expanded across disciplines. The scope and range of analysis in energy-related research broadened significantly with the introduction of a critical perspective on energy consumption and the entrance of previously absent disciplines into scientific debate (*Weber*). Widening critique of the idea of "progress" and the implications of new technologies and discoveries have made ethics as a new topic in many disciplines.

Complexity

Complexity is another keyword of the discourse, for several reasons. To begin with, the problems of society are increasingly complex and interdependent. They are not isolated to particular disciplines or sectors, and they are not predictable. They are emergent phenomena with non-linear dynamics. Effects have feedbacks to causes, uncertainties will continue to arise, and unexpected results will occur. With increasing complexity, problems of today's society have a self-referential nature, opening the door to paradox (*Goorhuis, Egger and Jungmeier*). The "binomial complexity-transdisciplinarity" relationship means that the reality being investigated is a nexus of interrelated phenomena that are not reducible to a single dimension (*Caetano, Curdao, and Jacquinet*).

Complexity occurs across contexts. Environmental problems comprise several sub-problems that fall into the domains of different disciplines. Hence, the research field is open and ill-defined (*Sheringer, Jaeger, and Esfeld*). The energy-saving problem has economic, technological, political, cultural, psychological, and social dimensions. In ski tourism, there is a double deadlock between ecological aims of conservation and economic development and between local and regional interests versus interests of the state (*Stalder*). In the aerospace industry, one of many industrial and commercial contexts of complex systems thinking, non-linear interactions between individuals or micro elements lead to symmetry breaking. Qualitatively different categories of phenomena are interacting, driving a qualitative restructuring that is cross-disciplinary in nature (*Jeffrey et al.*).

Complexity is also linked with another historical development. Technology assessment, *Nentwich et al.* observed, was a modernist concept. Transdisciplinarity moves beyond modernist paradigms and, thus, shares with the idea of "post-normal science" a claim that knowledge production is related to "unstructured" problems. These problems are driven by complex cause-effect relationships that demand increased collaboration with stakeholders in society (*Truffer et al., Chapter 5.2.2*). "Post-normal science" is also associated with issue-driven research characterized by a high divergence of values and factual knowledge in a context of intense political pressure, high decision stakes and uncertainties in epistemological and ethical systems (*van de Kerhof and Hisschemöller; Klabbers*). Finally, in the work of Basarab Nicolescu and "Manifesto: A Broader of Trandisciplinarity" in this book (*Chapter 4.3*), the new worldview of science furnished by complexity is fundamental to epistemological concept. The human being, in all levels of reality, is placed at the center of transdisciplinarity. Ethics, spirituality, and creativity are included in a new dialogue of culture and epistemology.

An International State of the Art

Europeans dominated the conference, in part because of the original purpose, to mark the conclusion of the SPPE. Yet, about fifty nations were represented among the nearly 800 attendees. The knowledge base is international in two respects – inter-European and North-South.

Inter-European

Political change has been a factor in expansion of the discourse in Europe. The fall of the Iron Curtain resulted in new initiatives in central Europe. Previously, the academic system of state-run research institutes was dominated by disciplinary specialties or narrow fields. Fundamental science and applied research were separated, and international access was limited. Legal barriers for creating independent research institutions have disappeared. New Centers for Energy Efficiency are platforms for disciplinary experts to work with local and regional administrations, governments, international institutions, and other NGO's in resolving urgent energy problems (*Gritsevich*).

Other European countries were also affected. In eastern Austria, regional development was forced to change. New environmental and sustainability measures have emerged in the multicultural region of the Central Danube. The project "BRIDGE Lifeline Danube" is a cooperative spatial planning process for development and a communication network among cities along the Danube. A newly created Environmental Research Network is now extending an earlier program of environmental research, dominated by natural sciences, into social sciences and education (*Kvarda; Katzmann*).

The discourse is also expanding as a result of comparative research. A project on dimensions of public concern about biotechnology involved fifty researchers from seventeen European countries. The empirical data covers policy process, media coverage, and perceptions of populations in each country. The study also sheds light on the dynamics of transdisciplinary research and value of discursive participatory procedures (*Dahinden, Bonfadelli, and Leonarz*). EUROpTA was an international comparative project on participatory forms of carrying out technology assessment that involved researchers and practitioners from Denmark, The Netherlands, Germany, Switzerland, the UK, and Austria (*Chapter 5.3.6*). The DACH project is an explicit effort to study strategies and control procedures in inter- and transdisciplinary projects in three countries – Germany, Austria, and Switzerland. Results illuminate the entire range of dynamics, from preconditions to criteria of evaluation (*Defila et al.* in *M20*).

North-South

"Syndromes of global change" are aggravated in the South by poverty and weaknesses in infrastructures (*Hurni*). The lack of traditions for inter- and transdisciplinary research and education is a further hindrance. Yet, expressing a belief voiced by many conference participants, editors of the forthcoming Encyclopedia of Life Support Systems (2000) declare that transdisciplinary learning and research first occurred in developing countries. This emergence was prompted by an urgent need to shift from ineffectual technology transfer to cooperation in development activities. Lay knowledge is also a strong dynamic in the South, where easily accessible forms of traditional technology and indigenous knowledge are valued. In the past, interventions were often conventional, one-way applications of knowledge that were not appropriate to local social, cultural, economic, and ecological realities.

North/South partnerships, the focus of many conference presentations and discussions, are a major platform for new transdisciplinary approaches. Partnerships have a long-standing tradition in participatory approaches and joint implementation. Swiss partnerships in Kenya and Madagascar, for instance, date from the 1970s. The changes that gave rise to the new discourse have resulted in new emphases. In 1997, the Indo-Swiss Collaboration in Biotechnology was restructured to respond to the changing socioeconomic status of biotechnology in both countries (*Jenny and Baumann*). In 1997, the Laikipia Research Program in Kenya was transformed into a research institution at the interface between society and academic science. Its spatial scope was expanded and new emphasis placed on scientific training, development practice training, and transfer (*Swiss Commission II; Künzi*).

Partnerships underscore the potential for a truly international knowledge base. The China Energy Technology Program is a cooperative effort of ABB Corporate Research, the Alliance for Global Sustainability (AGS), noted universities, and Chinese institutions. A two-year program in Shandong Province is a test case for developing a globally applicable, generalizable methodology for analyzing the impact of electric power generation. The Chinese partner, a pivotal stakeholder and main data provider, plays a decisive role (*Bill*). The AGS is equally noteworthy. AGS is an international partnership that unites top research universities in Europe, Asia and North American with leaders from global business and industry, government agencies, and NGO's (*Kaufmann and Baud*).

The Global Need for Transdisciplinarity

Powerful lessons emerge from the comparative perspective furnished by an international gathering. Shared values, ideas, and approaches enable us to speak of a common discourse. Democratization and indigenization are points of connection across diverse problem and global contexts, from involving local citizens in solving neighborhood deterioration in Kongens Englave, Denmark to using indigneous Ghandian concepts of Swadeshi, Trusteeship, and the cultural model of a Nine-Square Mandala in a project on farmers' perceptions of livelihood security in India (*Hiremath and Raju*). National differences, though, remain important. *Nentwich et al.* cite the example of participatory technology assessment. Inevitably, it will differ from country to country. Moreover, successes in one country do not transfer automatically to another. Contexts of research and problem solving will also continue to differ between North and South. Yet, common factors characterize a transdisciplinary approach across countries (*Chapter 5.3.6*).

Transdisciplinarity, it should also be realized, is evident in countries that were underrepresented in Zurich. To take the US as an example, several Americans presented sustainability and participatory projects in areas such as agricultural and farming research. They were too few to draw general conclusions about the U.S. Yet, "transdisciplinary" is a label on a host of Internet sites. Examples range across learning assessment, arts education, distance education and special education; plus mental health, the handicapped and children with multiple disabilities, rehabilitation and pain management. The term also appears on web sites dedicated to engineering problems, ecological economics, human population biology, language and thought, preparation for teamwork and collaboration, cybernetics and infomatics, and knowledge organization. More recently, the National Cancer Institute and the National Institute on Drug Abuse joined with the Robert Wood Johnson Foundation to fund new Transdisciplinary Tobacco Use Research Centers. Not all Centers engage collaborative research with community stakeholders, but collaborative problem-focused research with multiple partners is a primary value in this new initiative.

To speak of a shared discourse among the heterogeneous group of people who gathered in Zurich for the Transdisciplinarity Conference is not to demand conformity to a single definition but, to echo *Bernard Giovannini*, acknowledge a "general goal." There is no party line, or, an image that arose in the final panel of the conference, a "church" of believers (*Chapter 6.3*). Industrialists interested in more effective product innovation, through user feedback in competitive markets, sat alongside academics interested in critique of science and the market economy.

What put them in the same room was a common realization – they need each other. Some proclaim the arrival of a new paradigm, meta-theory, science, or discipline. The majority, though, regard transdisciplinarity as a complement to existing disciplines, an essential cross-disciplinary methodology for creating insights and solutions, a means not an end. *Mey and Kapila*, in *Chapter 6.3* sum up the conviction of many: transdisciplinarity is a major tool for reaching a difficult goal, sustainable life. Transdisciplinarity, however, is not a mere handmaiden. It is neither a "new science" nor merely a "sideline of a "by-product" (*Scheringer, Jaeger, and Esfeld*). Existing and new approaches are combined in a collaborative effort to create new spaces and cultures of mutuality.

3

Keynote Addresses

3.1 What Kind of Science Does our World Need Today and Tomorrow? A New Contract between Science and Society

by Charles Kleiber, State Secretary for Science, Berne, Switzerland

Abstract
The progress of science and technology is great, but so are challenges that confront us. They raise fundamental questions about the purpose of human existence, our relationship to nature, and the kind of science we need. Transdisciplinarity is an integral part of any response. A new collective intelligence and creative stewardship are required, capable of overcoming the territorial boundaries that currently limit higher education and research and forging a new relationship of three major forces – science, democracy, and the marketplace. They will also require incentives, new initiatives, and high standards of quality.

Introduction

The history of humankind during the last century is marked by massacres of unprecedented scope and barbarity. We say farewell to this ominous chapter in our history by counting our dead. Ravensbrück, Dachau, Guernica, Hiroshima, Phnom-Penh, Kosovo, and Grozny. Examples of the defeat of reason and democracy such as these are seared deeply into our collective consciousness.

- Poverty, starvation, low levels of literacy, poor standards of living and dismal life expectancy: a growing gap separates us from the South.
- Pollution, climatic disorders, the poisoning of our rivers, the uncontrolled growth of many of our cities, desertification, wasted water resources: these countless assaults and threats have had a dramatic impact upon our beautiful and still unique planet.
- Mad cow disease, dioxin poisoning, HIV/AIDS-contaminated blood, and ordinary deaths on ordinary roads: these unexceptional tragedies accompany us into the new century.

All of this constitutes a bleak inventory indeed.

But this bleak tableau has also been matched by the most amazing discoveries imaginable. In the last century we penetrated into the innermost realms of matter and life itself. We ventured into parts of the universe beyond our solar system. We explored human consciousness beyond words and their illusions. In a short span of time, our understanding of the world, the universe, and ourselves have undergone a profound transformation.

What an extraordinary inventory of our achievements! Cell phones, the Internet, the ability to gain access to all the memory and all the knowledge in the world. Through increases in mobility, incredible gains in productivity, as well as an increase in our potential for total destruction, we plunge into the new century lost in wonderment, simultaneously marveling and worrying about transformations that have been too sudden to be controlled, too rapid to be thought through. One thing is clear: our inability to think through these transformations feeds both our hopes and our fears.

We stand ambivalent before a tempered inventory, bathed in chiaroscuro.

No other century has awakened so much shame, bitterness, confusion, wonder, hope and fear in the hearts of humankind. Never has the progress of science aroused so much promise, so much doubt. It is as if knowledge, science, and technology have engendered more questions than answers. Vital questions abound. What is the purpose of life? What in fact do we gain by doing everything faster and quicker? To what ends must we put our newly won autonomy? Is it possible to live in harmony with nature? Put simply: *why* has not vanished even as we tackle *how*.

Moreover, the very status of truth has changed: henceforth, there are no longer stable and ultimate truths, only partial, fragile, and often contradictory ones. Stripped of these comfortable certainties, each of us is faced with the responsibility of building his or her own set of values and meanings. More than ever, the answers to questions about the meaning of life and about the purpose of one's own existence must be personal and individual.

In light of all this, it is crucial to ponder the question "What kind of science does our world need today and tomorrow?" And in order to reflect on this lofty theme, we must begin with both an honest assessment of our past tragedies and a critical appraisal of the fabulous promises made by the future.

It is essential, first of all, that each answer be rooted in living history. Each premise has to take into account the interrogations of the men and women, our brothers and sisters, who, like us, have been collectively and individually responsible, and who have chosen to act for the better and the worse.

The answers to the question "What kind of science do we need?" will depend, I believe, on three major forces that progressively took shape in the last century and that will continue to decisively shape us and our societies: democracy, science, and the market economy. These three mutually dependent forces, which also represent three contradictory systems, have come to constitute a *ménage à trois*, so to speak. Theirs is a *ménage* characterized by conflict, at times rife with tension, but nonetheless capable of reconciliation and working together. Indeed, a creative and civilizing relationship can emerge when democracy and politics are able to carry out their mission of social integration and to use science in the best interests of humankind. But, its potential as a destructive and merciless *ménage* is evident when a matrix of common values does not bless its union and the desire of its members to live together.

Using these three historical forces – democracy, science and the market place – as a basis for my discussion, I will now undertake to define the mechanisms capable of liberating scientific creativity. The rest of this chapter is a reflection on the body of values needed to guide scientific development and help science measure up to the great challenges that lie ahead of us. We shall see too that transdisciplinarity is an integral part of any reflection about how to meet these challenges.

Beliefs Underpinning My Reflections

Three strongly held beliefs underpin my reflections in this area, and before proceeding further I would like to share them:

- First, I am convinced that despite all our worries and doubts, it is possible, not to mention urgent, to set in motion a process of cultural, political, social and economic innovation. This process must of course be based on the opportunities, promises, and historical perspectives available to us today, but it will be capable of inventing the future by generating *une intelligence collective*, that is knowledge shared by all.
- Second, I hope to show that in this innovative process, higher education and research can play an essential role by providing the specialized sciences and skills without which knowledge cannot progress. But, they must also contribute to this process by throwing into question the territorial boundaries of knowledge that scientists diligently set in

place in the last century. Scientists constructed these boundaries not only in order to progress, but also to state their identity, protect their autonomy, and wield power. Today it is vital to recognize that the age of knowledge as a thing to be possessed is drawing to a close, and that we are rapidly entering the age of knowledge as a thing to be passed on. *C'est le temps des passeurs, après le temps des propriétaires.*

- Third, I believe that our country has everything needed – creativity and wisdom, imagination and pragmatism – to set up a new contract among democracy, the marketplace, and science. This new contract should allow us to control the pockets of hate that exist in every society and to help each member of society truly accept the Other. In my view, this then is the definition of the *mission civilatrice* of democracy and science.

Democracy, The Market Economy and Science

It is a fact that the globalization of the free market economy, with its celebration of the individual, generates social atomization and a concomitant loss of shared values. Furthermore, since the fall of the Berlin Wall in 1989, the market economy has had no real rivals. It is of course also important to acknowledge that the market economy is based on a principal contradiction, succinctly stated in 1705 in Bernard de Mandeville's aphorism "private vice produces public good." Without a doubt, the market economy has the potential to transform personal interests into good for all. But fabulously powerful personal interests may in the long run also destroy the market economy if they are not held in check by community values. *Homo economicus* may be free but is also alone, devoid of ties, and with neither history nor shared beliefs. This, then, is the risk of short-term logic of profit, and that is precisely why the market economy must be integrated back into culture and society.

How are we to go about the painstaking, but necessary, construction of a new international public order capable of integrating market forces and community values? This effort requires the invention and implementation of new instruments of regulation at the global level. These new instruments might include, for instance, new jurisdiction governing the use of the Internet and new jurisdiction on pollution, crime, commerce, and intellectual property, to name only a few examples. The effort also requires that we develop new forms of solidarity which reflect our increasing economic interdependency. Finally, it is essential that we invent new rules favorable to the development of the South. It is simply inconceivable that the present North-South gaps be allowed to continue in the long run. Resolutions such as those outlined above constitute, I believe, the long-term logic of common good and shared values.

It is relevant at this point to address the issue of the acceleration of scientific and technological innovation. At present, science and technology have succeeded in being able to produce more useful goods, a longer life, and more comfort and well-being – all by controlling nature. Moreover, we are only at the beginning of exploring these new frontiers. It is clear, too, that competition in the areas of science and technology will continue to increase, even to the point of triggering different forms of struggle. The French biologist Lwoff aptly identified one of the realities of scientific research when he stated, "Science, the work of man, is inhuman, absolutely so." Scientists are scientists, not ethicists, and the pressures placed on them to go further and further in the adventure we call knowledge will be enormous. Finally, if we reflect on Gabor's prediction that "everything technically feasible will be undertaken; everything that can be sold will be manufactured," we may begin to realize the degree to which science and technology exist both as promise and as threat to our society.

Having awakened from the dream of the eighteenth-century French encyclopedists, we now recognize that it is of course impossible to command a totalizing knowledge of everything and to understand the world in its entirety. Indeed, it might be said that today, in a full pendulum swing, we inhabit a new tower of Babel. We live in a time of dispersed knowledge, with each domain well defended by its professional guardians. The effect of this compartmentalization is that dialogue has become impossible. We might even say that civilization, as we know it, is being thrown into question. Knowledge gained in one sphere – democracy, the sciences, or the marketplace – is not being transferred to the others, and a poorly informed elite, consisting of uncultivated scientists and uninformed leaders, constitutes a hindrance to any viable transmission. That is precisely why it is so crucial to convert implicit knowledge into explicit knowledge based on common criteria. In other words, it is crucial to reconcile popular, traditional, academic, and technical forms of knowledge. Again, it cannot be emphasized enough that it is time for messengers capable and desirous of conveying knowledge. In short, it is time for true transdisciplinary collaboration.

How does this change in the nature of knowledge come about in society? It takes the form of a continuous process of transformation based on an exponential increase in information exchange. This exchange is more and more powerful and less and less costly. The end result is that "The Great Revolution" is now a thing of the past. In its place, a multitude of silent revolutions greet us each day. Henceforth, we are involved in an interactive process with regard to our world: we make the tools that in turn make us. The most representative example of this process is of course the Internet. What is certain is that a new attitude to change is a prerequisite to accepting the challenges that now face us. It is no longer possible to effect change by decree. Change must be bought, measured, and computed, from now on.

Paradigm Shifts

Let us recall, schematically and as a preliminary step, a few of the epochal paradigm shifts that society as a whole and the scientific community, in particular, have undergone.

- First, we no longer view knowledge as a gift from God but increasingly as a factor of production.
- Second, we are also rapidly evolving from an economy of raw materials to one of gray matter.
- Third, where once we believed nature was immutable and perfect, we have now come to accept the fact that it is a constant mutation.
- Fourth, science, formerly considered autonomous and unified, is today seen as accountable to society and fragmented.
- Fifth, education, previously considered a personal privilege, is now recognized as a social, political and economic requirement.
- Sixth, science, long believed to be the source of truth, is now seen as a source of much uncertainty in our values and meaning.
- Seventh, knowledge as an instrument of social reproduction and of individual recognition has been transformed into knowledge as an instrument of social innovation and collective intelligence.

These are fundamental and far-reaching changes in our relationship to knowledge and learning. These paradigm shifts have also brought with them a certain number of both conflicting and convergent trends, which this chapter can only enumerate.

I suggest we begin by identifying the conflicting trends, presented here in binary pairs: profit vs utility, profit vs security, as well as profit vs truth. Conditions favoring the economy are also pitted against conditions favoring the common good and the short term against the long.

The ownership of knowledge, too, stands in marked opposition to the sharing of knowledge; and democratic systems based on opinion contrast sharply with those based on knowledge. With the context of this particular forum, it is easy to understand how, in our final binary pair, disciplinary obstinacy intrinsically conflicts with transdisciplinary passion.

Certain trends, on the other hand, converge. This development may be illustrated by the following coupled pairs: the struggle to conquer poverty converges with the sharing of wealth; the struggle to reduce ignorance is in essence a sharing of knowledge; the fight to eradicate disease converges with efforts to provide better health care for all; efficient infrastructures are the natural partners of sharing information and circulating goods; economic growth favors the sharing of purchasing power; and finally, scientific and technological conquests converge with efforts to control nature for the good of humankind.

Even a cursory overview of these different trends foregrounds the need for reconciliation, a subject I will return to later.

Our goals are set on achieving bigger things, going further and penetrating deeper. At what price, however, will we be able to achieve these ends? To close this chapter and brief history of this *ménage à trois*, I would like, by means of the following quotation from Jean-Pierre Dupuy, to capture the essence of the driving force that, for better or worse, creates the power of democracy, the market economy and science:

> *On ne saurait avoir assez d'une bonne chose. Ce qui compte, c'est moins ce que l'on a en soi que ce que l'on a par rapport à ce qu'il serait mieux d'avoir. Or ce qu'il serait mieux d'avoir croît avec ce que l'on a et avec ce que les autres ont. D'où cette surenchère sans fin, l'abondance créant autant de raretés qu'elle en supprime.*
>
> *(One cannot get enough of a good thing. What we possess counts less than what it would be better to have. However, the desire to possess increases with what we, and others, already have. The result is a never-ending escalation, where abundance generates as much scarcity as it eliminates.)*

From Maurice Bellet we also have this disquieting commentary: "Peace of the soul is forbidden, because it would break the engine of expansion." In this *ménage à trois*, all three forces feed on the possession of objects, both material and symbolic. This is true of artists and scientists alike.

A Call for a New Kind of Knowledge

In a game that knows no boundaries – a game that contaminates science, democracy and the market economy – how can we distinguish true needs from simple whims of fashion? How can we distinguish between necessity and fancy? How can we differentiate conviction from opinion? What is the meaning of this all? Where is the civilizing project? Where is the universal outlook of the minds that might be capable of counteracting the global reach of the market? Where is the common ground that links each of us to the other?

It is this missing link that feeds modern disenchantment with the world, wariness towards anything new, and the distrust that endangers our future. What kind of science do we need today and tomorrow? We need the kind of science that can live up to this need for universality, the kind of science that can answer these questions. On this point, I would like to share with you some of my reflections.

In answer to the question of what kind of education and science we need, I suggest we need a new kind of knowledge, a new awareness that can bring about the creative destruction of certainties. Old ideas, dogmas, outdated paradigms, moldy knowledge, and decayed know-how must be destroyed in order to build new knowledge of a type that is more socially robust, more scientifically reliable and stable, and above all better able to express our needs, values and dreams. What is more, this new kind of knowledge, which will be challenged in turn by ideas yet to come, will prove its true worth by demonstrating its capacity to engage in dialogue with these ideas and grow with them.

My concern here might also be formulated in the following way. Can we locate the fragile and unstable equilibrium where democracy, the marketplace, education and science acknowledge each other, strengthen each other, and yet hold each other in check? This point of equilibrium – if indeed it exists – must be the result of a collective effort fueled by a respect for common goals. It must be the result of a sustained quest in which transdisciplinarity will play an essential role. This brings me directly to the theme that is the focus of this forum.

Why is Transdisciplinarity Necessary?

Why should disciplines be brought together? It is becoming increasingly clear that all three historical forces – the market economy, science and democracy – not only benefit from, but more importantly require, working across disciplines.

- First and foremost, the market economy requires polydisciplinarity. The complexity of today's world demands that any given participant be able to combine the responsibilities and talents of multiple roles. We are called upon to be a new breed of hybrid professionals, becoming, for example, both economists and lawyers at the same time, engineer-economists as well as economist-architects.
- Second, science itself requires interdisciplinarity. How else can we study the effect of climatic disorder on genomics, if not by adopting a transdisciplinary approach?
- Third, democracy requires transdisciplinarity. Indeed, it is a *sine qua non* condition underpinning the proper functioning of the *agora* in a complex world.

Now, it is obvious that a number of disciplines have slowly begun to converge. There is a great danger, however, that this convergence will be a process of simple juxtaposition. Our objective must be to combine knowledge from different fields and traditions in such a way as to increase their power of expression and interpretation. We must strive to achieve the relevant integration of the differences that each field brings to the new union in order to achieve true transdisciplinarity. This is the reason why intellectual interference is our duty, why we have what could be expressed in French as *le devoir d'ingérence intellectuelle*. It is imperative that we be concerned with questions that do not seem to concern anyone else, with questions, so to speak, "that belong to nobody."

In the move towards transdisciplinarity, one vital question must be addressed. How can we go about combining disciplines without losing the strengths of each discipline? To begin with, it is fundamental that certain things be reconciled. We will need, for instance, to reconcile the logic of academia with that of stewardship. Different traditions of knowledge will need to be reconciled as well. Furthermore, traditionally opposed local and global approaches, championed by "globalomaniacs" and "globalophobics" respectively, will need to find compromises that allow them to give the best of themselves. In a similar fashion, the job market and the education market will have to collaborate in order to create an active interface between the world of labor and the world of education. Last, but certainly not least, scientists and poets will have to come together and engage in fruitful dialogue across disciplines. This, in my opinion, is one of the most strategic points of reconciliation. It involves not only being more accepting with regard to *la pensée critique*. It also provides a vital context in which to create new metaphors and new myths able to articulate our new world.

Apart from the reconciliation between these different entities, combining disciplines requires that we concern ourselves with managing collective memory in a transdisciplinary fashion. This entails creating a dynamic dialogical movement between the past, the present and, most importantly, the future.

Additionally, the successful combination of disciplines can only take place by stimulating genuine multiculturalism. Token gestures in this regard are insufficient. Switzerland must take great care to truly exploit its immense resources in this area, as they constitute one of its greatest assets.

Furthermore, transdisciplinary success requires that we promote the dialogue between science and the *cité*, without which there can be no science, just as without science there might be no *cité*. On this point, it is well to listen to Sir Bertrand Russell when he points out that science made a God of us before we learned to be men and women, and to remind ourselves that only by attending to society and its needs scientists will be able to remain within the community of their fellow human beings. Indeed, one of the fundamental functions of transdisciplinarity has to be the formulation of questions that rise from the *cité*, to grasp and give voice to the doubts of the *cité* and of society as a whole. So, universities must give priority to questions rather than answers, and academia will have to facilitate the generation of new, pertinent questions rather than churn out old answers that simply respond to obsolete questions.

In order to stimulate transdisciplinarity we also need strong incentives and new initiatives. For example, it is worth considering that a given percentage of subsidies allocated to research should be dedicated to transdisciplinarity, as suggested by certain members of the scientific community. It is crucial, too, to foster the development of and give incentives to transdisciplinary centers of research, *Wissenschaftkolleg* such as the ones in Zurich (Collegium), Berlin, Budapest, Bucharest, and those in the United States. We must give priority to what makes transdisciplinarity possible: the pooling of disciplinary knowledge and information, technological revolutions, and the creation of networks and new forms of knowledge.

What is more, transdisciplinarity cannot be developed in a vacuum. Our first duty must be to closely examine the results. Consequently, rigor and vigilance are important criteria in this undertaking, and the implementation of high standards of quality and scientific validity is essential.

These standards of course will have to be invented, established, and enforced. However – and this is of vital importance – at the very same time, there must be a certain amount of generosity and openness with regard to the early, fledgling results. Exceptional care must be taken to avoid destroying the momentum and the enthusiasm that nourish the early results of transdisciplinary work.

Finally, the foundation of this edifice is the education of men and women, regardless of age, who are technically competent and who are able to learn how to learn, to adapt, to engage in teamwork (yet another example of collective intelligence), to communicate, to lead and, most important of all, to be autonomous. This is crucial because it represents a genuine learning revolution. It can start, though, exactly as it is starting here, with a project such as the one in which we are currently engaged. Learning must become an *art de vivre. La curiosité, ce feu*, this flame, as Montaigne terms it, must light the way.

Conclusion

To conclude, I would like to raise the question of what this small country can contribute to the agenda of a new contract between democracy, science and the marketplace. In order to illustrate my point, I will begin by evoking the image of Swiss democracy – the product of the unwavering will of its founders – as no less than the political embodiment of the virtue of patience. Those who dare to undertake a life in politics, those in particular who are filled with *ardente patience, mit brennender Geduld*, are in many respects modern-day saints. Need I say that we are lucky to have among us a few of them. There is no doubt whatsoever that we will need their help to transform our system of research and higher education.

Swiss democracy, this state of will, is fearful of conflict and thus not accustomed to using its conciliatory forces. Although at times it goes unrecognized, the advantages of this fear of conflict are nevertheless significant. The Other is always present in the political debate. The adage, *Il n'y a pas de fatalité à s'entendre* (getting along is not inevitable) expresses our situation well, and aptly speaks to the foundation of our culture of tolerance. This capacity for tolerance will allow us to add to our remarkable scientific heritage – the new treasure of transdisciplinarity.

Conflict, or better yet debate, is necessary. Reason is not a quiet stream, but a tumultuous river, and diving into it entails risk. The quest for truth necessarily pits itself against conventions, *la pensée unique* or political correctness. That is why this joint reflection on transdisciplinarity is necessary and why I would like to extend my gratitude and appreciation to its organizers for their laudable work.

I have undertaken to outline utopic visions that are nonetheless realizable. It is now time to work together to negotiate and to demonstrate a wise pragmatism rooted in constructive knowledge. Everything is possible. Hope, of all kind, is very much alive. Yet, we do not have much time. We must heed the warning of General Kotousov, the victor of the battle of Russia: "There is no problem that a lack of solutions cannot solve."

If we want to maintain and even improve our positions in the transformation of knowledge and the world that is currently underway, we will have to work hard. Lest we doubt, let us remind ourselves that transformation and change will not destroy us. On the contrary, to change is to be faithful, that is to say to be faithful to the essential. I would like to close with a thought by André Bonnard, which goes to the very heart of the matter:

> *Connaître c'est échapper à la solitude, c'est participer à la vie d'autrui et du monde. Et si cette connaissance est en nous non point inerte mais active, génératrice de valeurs, connaître c'est participer à la recréation de nous-mêmes et du monde.*
>
> *(Knowledge is a way of escaping solitude, of sharing in the lives of others and the world. And if interior knowledge is not lifeless but is instead alive, giving birth to new values, then knowledge becomes a means of participating in a recreation of ourselves and of the world.)*

3.2 From Microscope to Kaleidoscope: Merging Fields of Vision

by Rita Colwell, Director, U.S. National Science Foundation (US NSF)
Delivered by Robert Eisenstein, Assistant Director for Mathematical and Physical Sciences (US NSF), Arlington, Virginia, USA

Abstract

Science and engineering are on threshold of a new connective power. Invoking the metaphor of a kaleidoscope, new vistas are being revealed by information technologies and the science of complexity. Evidence appears at all scales in visualizations of a spiral, from a hurricane to gravitational waves. Fractal branching is evident in a river network, the veins of a leaf, and the propagation of cracks in fatigued materials. From the micro-level of an atom to the distant scale of astronomy, we are understanding the universe in new ways.

We are on the threshold of forging a new unity in science and engineering, a process that will drive our progress more than ever before. I would like to take us on a short journey to glimpse some of these zones of great creativity. We are currently celebrating our 50th anniversary at the National Science Foundation. I know that the Swiss National Science Foundation is looking forward to its 50th anniversary in two years, and we'll be congratulating you then. Scientific achievements over the past half-century have a historical heritage common to many nations and to all the disciplines. Our journey now brings us to the brink of unimaginable opportunity and discovery.

In the United States, we're very pleased that President Clinton has proposed a $ 675 million increase for the National Science Foundation for this coming year. That is double the largest dollar increase ever proposed before. It is a jump of 17.3 percent. Our budget would increase to about $ 4.6 billion. We cannot imagine a better anniversary present. In fact, we highly recommend this as a way to celebrate a fifty-year anniversary!

This budget will help us meet the most critical challenge: to strengthen the core science and engineering disciplines while fostering and nurturing links between fields and preparing the next generation. Our philosophy at NSF is also to invest in those initiatives that will generate benefits across the spectrum of science, engineering, and society. Today I would like to explore a new kind of vision that encompasses the breadth of our mission at NSF. To that end, we have titled our talk, "From Microscope to Kaleidoscope: Merging Fields of Vision."

From Microscope to Kaleidoscope

The microscope represents an approach that much of modern science has followed up to now. This approach seeks understanding by taking things apart into their components. It has given us the lion's share of scientific knowledge to date. The microscope itself is a tool that revealed fantastic new worlds, each more minute and requiring finer focus than the last. It has been almost exclusively a reductionist approach. But as science and engineering grow ever more integrated, we need a new metaphor. I've chosen the kaleidoscope. The word itself is fascinating: it derives from Greek, and roughly means "beautiful form watcher." When we turn the kaleidoscope tube to create shifting shapes and colors, we watch new and unpredictable patterns and hues appear. At the same time, the elemental components retain their integrity. So it is with the scientific disciplines, which retain their unique viewpoints and insights, but intersect increasingly with other fields to forge new frontiers at every scale.

The American naturalist and writer, John Muir, saw an essential truth mirrored in the natural world. His words were: "When we try to pick out anything by itself, we find it hitched to everything else in the universe." Only through mapping and nourishing these linkages can we truly reflect and probe the wholeness of the world that we study. We would like to explore how integrating the disciplines of science and engineering will let us see into the kaleidoscope of complexity in all its beauty, dynamism, and unexpectedness.

New Vistas of Complexity

Our new vistas, our new ways of seeing, are provided by an array of new vision-enhancing tools. Among the most prominent are information technology and nanotechnology. In this dynamic vision, our tools are actually accelerating the merger of the disciplines. The entire enterprise must progress as a whole; gone are the days when a discipline could go it alone.

We now begin to see that many phenomena and processes in a variety of fields are uncertain, even unpredictable. For the first time, the embryonic science of complexity lets us begin to study these processes. It's an oft-stated truism that the whole is sometimes much more than the sum of its parts. In science, we are increasingly aware that new emergent behaviors can arise when one combines individual units to create a larger whole. Whether one studies the behavior of a zebra herd, or tries to model business success in a world economy, the science of complexity is an increasingly fertile way to begin. Indeed, we can trace this principle across the disciplines, from economics and sociology to ecology and physics.

Why do we care about complexity? Because it can give us a perspective spanning all fields of study and all scales. We can symbolize this with an example that connects across truly vast dimensions. Visualize the form of a spiral, and imagine it taking shape in a hurricane – perhaps one the size of our state of Florida – as viewed from a satellite. Then leap much farther out into space and see the same form in a spiral galaxy that is 100,000 light-years across. Finally on the grandest scale we can imagine gravitational waves rippling across the largest scales in the universe.

In fact, the National Science Foundation has just inaugurated an observatory called LIGO whose ultimate goal is to detect gravitational waves. LIGO stands for the Laser Interferometer Gravitational Wave Observatory. I'm very pleased to say that LIGO will be a part of an international network of such detectors, with other sites existing in Italy, in Germany, in Japan, and possibly in Australia. It's also particularly fitting for this audience that the prediction of gravitational radiation was one of the most profound pronouncements of your most notable alumnus, Albert Einstein.

Biocomplexity

Deep common principles underlie both the physical and the biological sciences. For example, we can see the same principle of fractal branching at work, whether in the form of a river network, in the veins of the leaf, or in the propagation of cracks in materials that are fatigued. We at NSF have framed a new and encompassing approach to studying our world. Our term for it is biocomplexity, and it is one of our key budget emphases at the National Science Foundation.

"Biocomplexity" is an interdisciplinary view of the complex interactions in biological systems, including humans, and between these systems

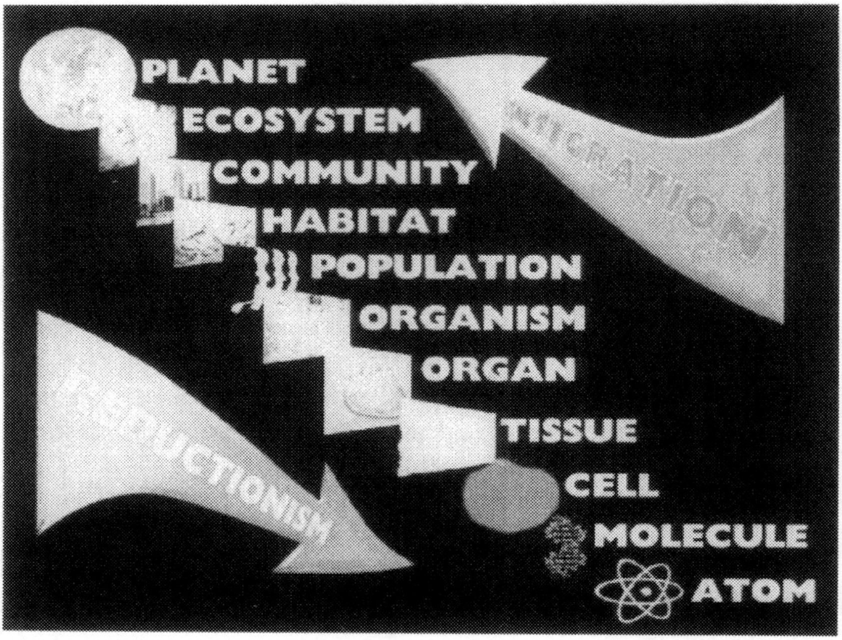

Figure 1
Biocomplexity: It is vital to understand the complex relationships between living organisms, including humans, and their environment. When we use the concept of complexity to look at life across varying scales, our living world becomes more than the sum of its parts.

and their physical environments (Fig. 1). We are ready for this approach because we're on the brink of developing the tools to observe complexity at multiple scales, with nanotechnology, with global ecological observatories, and vantage points at scales in between. We know that ecosystems do not respond linearly to environmental change. Tracing the complexity of the earth's environment is profoundly important to the future of life on our planet.

An example of biocomplexity in practice illuminates this new perspective. To restore the Florida Everglades, it is essential to understand how different hydrologic schemes will affect key species. Researchers with support from NSF and the U.S. Geological Survey are developing models down to the level of individual animals in panther or deer populations. They are able to construct finely detailed maps that show how water releases will shape habitat quality for different species. Assembling this bigger picture, of course, takes tremendous computing power as well as insights from ecology, mathematics, economics, and sociology. The end result, however, is a practical tool for policy makers.

Human Contexts

If we move to another level and look at our own species, we see a system as complex as any, and the disciplines are converging to chart it. The cover story in an issue of *Science Magazine* last year featured the earliest evidence of the use of tools by our human ancestors, along with a reconstructed skull. In the great leap to our modern tools, another recent *Science* cover story showed a computer reconstruction of a brain that gives the endocranial capacity for an early hominid. Tracing how our brain evolved is a major scientific challenge.

We are also employing simulations of the brain in action, what are called "brain templates" – a technique that might eventually help point to sections of the brain that are active in the learning process. This is a focus of research we're very interested in for future emphasis at the National Science Foundation.

Speaking of learning, I would like to turn to a marvelous demonstration of the power of information technology for the classroom. Through the University of Illinois Bugscope project, schoolchildren from around the country get the chance to control a scanning electron microscope through the Internet to take snapshots of their favorite insects (Fig. 2). The results are fantastic and there for anyone to enjoy on the Bugscope website (<http://bugscope.beckman.uiuc.edu>). The students get a taste of what it is like to submit a proposal, with their teacher, and how to schedule time on a scientific instrument.

Connections are fascinatingly abundant between medicine and fundamental sciences as well. Magnetic Resonance Imaging (MRI), for example, has improved basic health care. It is a well-known and non-invasive technique that is used to diagnose many illnesses. What is less well-known, however, is that MRI springs from basic research in physics, chemistry, and mathematics.

Scales of Connection Making

Information technology illuminates other connections. For example, anthropologist Eric Delson of the City University of New York and colleagues define the dimensions of a primate's skull with three-dimension-

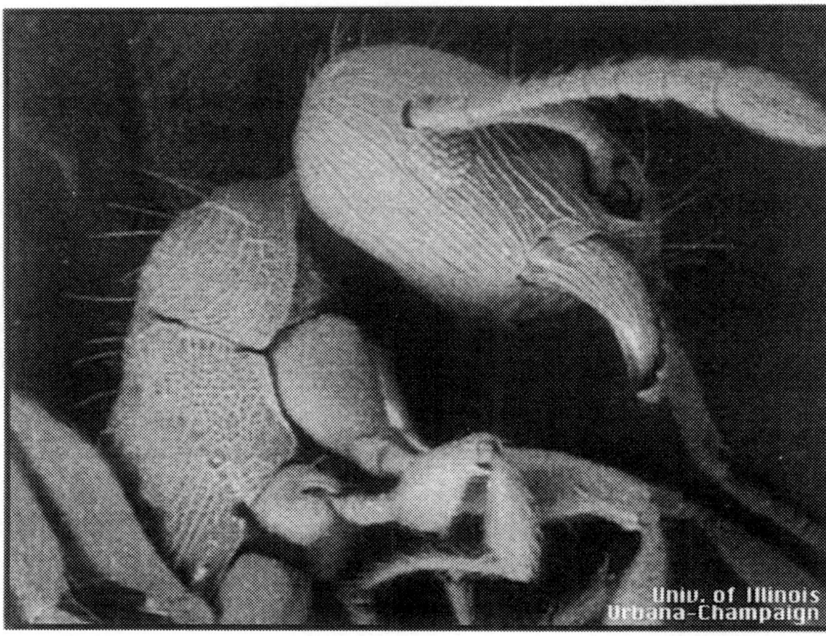

Figure 2
Bugscope. The University of Illinois has pioneered an effort to make a scanning electron microscope available through the Internet to school children across the United States and abroad, allowing them to participate in making highly magnified images of insects.

al coordinates, or landmarks. As they trace the differences between subspecies, species, and genera, they hope to uncover pathways of primate evolution, perhaps shedding light on our origins.

We can move to astronomy for another example of creative combinations. Just eight months ago we dedicated the Gemini Telescope in Hawaii, Gemini North, one of the new generation of ground-based telescopes with unprecedented reach. Gemini South, located in Chile, will be completed around the end of the year 2000. Fitted with adaptive optics, Gemini, an earthbound telescope, has been able to achieve the same resolution and clarity as the Hubble Space Telescope, but from the ground, in the infrared part of the spectrum (Fig. 3).

The technique has unexpected applications. The National Science Foundation has just launched a Center for Adaptive Optics at the University of California-Santa Cruz. This center has a dual mission: It will refine the technique not only for astronomy, but also for research on the human eye. Supporting these kinds of science and technology centers is

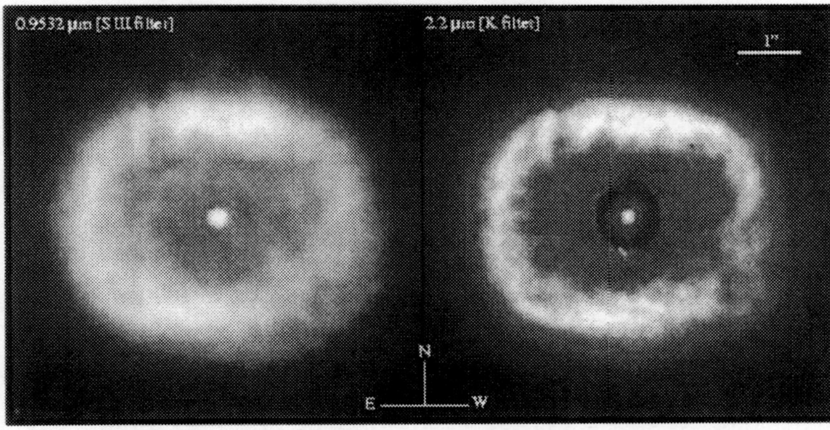

Figure 3
Nebula viewed from Earth and space. A planetary nebula is viewed by the Hubble Space Telescope, shown left, and by Gemini in the infrared, right. With adaptive optics, Gemini – an earthbound telescope – has been able to achieve the same resolution and clarity.

one way we can institutionally encourage the interchange between disciplines.

Returning to earth, we can trace the impact of our new tools on anthropology. Molecular genetics, for instance, has been applied to one-sixth of the world's population that married under the rules of the Hindu caste system for the last 3,000 years. Lynn Jorde of the University of Utah found that marriage patterns are mirrored in different groups' DNA sequences. These reflect that women can marry up in social rank, but that men of lower rank are restricted in such social mobility. Furthermore, the DNA of upper-caste individuals bolsters historical evidence that the caste system originally came into India from the northwest.

New Frontiers of Evidence

Sometimes visualization is just not enough; we need to hold the evidence in our hands, and so we turn to tele-manufacturing – what has been called a mix of science and sculpture. With this technique, a computer helps produce physical three-dimensional models of everything from a human heart to a landscape.

Information technology and complexity also give insights into the human heart, through computer simulations of the heart's electrical activity. James Keener, a University of Utah mathematician, and his colleagues have modeled a healthy heart and have gone on to model patterns typical of fibrillation, the uncoordinated electrical activity that often leads to heart attacks. They are investigating why some patterns of electrical stimulus are better at eliminating fibrillation. People with pacemakers and implantable defibrillators could benefit.

Moving to the atmospheric sciences, we find more evidence of the power of computing. Terascale computing gives us the ability to predict storms faster and on a finer scale, revealing violent storms that current prediction systems miss altogether. If we look at the scale of counties in Oklahoma, we find that the existing U.S. National Weather Service System can miss the presence of a tornadic storm, but that a prototype terascale system can predict the storm. Five years ago, this sort of prediction was deemed impossible, but with terascale systems it will be routine.

Finally, we can twist our kaleidoscope and take us to one more dimension to the Lilliputian level of the nanoscale. The National Science Foundation is now leading a multiagency initiative on nanoscale science and engineering. At the frontier of nanotechnology, we are beginning to manipulate matter at the scale of atoms. As it happens, one nanometer, one billionth of a meter, is a magical point on the dimensional scale. Nanostructures are at the confluence of the smallest of human-made devices and the large molecules of living systems. Red blood cells, for instance, have diameters spanning thousands of nanometers. Micro-electrical mechanical systems now approach this same scale. We are at the point of connecting machines to individual cells.

Today it is clear that every discipline and every scientist and engineer and educator adds a unique element to the ever-shifting kaleidoscope of discovery. As we watch our myriad fields of vision merge and produce new ways of seeing, we see no limit to expanding our vision. Indeed, we must broaden it, to a vision that pursues greater resources for all of science and engineering and deepening connections among our nations. There is no limit to what we can learn and do, if we work together as one community of scientists and engineers. It is a vision of synergy, the tracing of complexity, the propulsion of information technology, and the integration of all disciplines. This is a vision whose fulfillment will require the energy and talent of every one of us.

3.3 The Potential of Transdisciplinarity

A Joint Presentation by Michael Gibbons, Secretary General of the Association of Commonwealth Universities, London, UK and Helga Nowotny, Chair of Philosophy and Social Studies of Science, Swiss Federal Institute of Technology, Zurich

Abstract

A transformation is occurring in the relationship of science and society. A new mode of knowledge production is at the heart of this transformation. Much of the thrust of innovation is coming from new links between traditionally segmented producers and users of knowledge. Contextualization of research around the interests of stakeholders fosters a more "socially robust" knowledge that transgresses disciplinary and institutional boundaries. The ancient Greek *agora* is a model for a new transdisciplinary forum.

Introduction

Gibbons

The book referred to in the previous session today is *The New Production of Knowledge*, and, it is fair to say, the book has caused a modest stir in universities and some policy circles. I also think the work has been only superficially understood. So, Helga and I are going to try, during this joint presentation, to illuminate some of the points we regard as crucial to understanding what we were saying in *The New Production of Knowledge*, particularly as they bear on the issue of transdisciplinarity.

To sum up what has happened so far and to set the scene for what we want to do, four things this morning have struck me with particular force (with reference to the keynote speeches of *Charles Kleiber* in *Chapter 3.1* and *Rita Colwell/Robert Eisenstein* in *Chapter 3.2*, as well as the first panel reported by *Alain Bill, Sybille Oetliker, and Julie Thompson Klein* in *Chapter 2.1*).

- First, patience: If you are going to work in transdisciplinary research, you need to be very patient indeed. The evidence clearly shows that developing transdisciplinary teaching takes time and commitment from both academics and their institutions.
- Second, science and society are still being treated as unproblematic categories. In fact, they have both become problematic categories.
- Third, knowledge is transgressive. Nobody, in my awareness, has succeeded anywhere for very long in containing knowledge. It seeps through institutional structures like water though pores of a membrane. As with liquids in membranes, knowledge seeps in both directions, from science to society and from society to science.
- Fourth, the sub-theme of the conference – "joint problem solving": Rudolf Häberli and I chose that word one afternoon in London. We did so, because we wanted to get away from the idea that transdisciplinarity involves more than simple juxtaposition, more than laying one discipline alongside another. Rather, "joint problem solving" is intended to convey integration of perspectives in the identification, formulation, and resolution of problems.

With these four points in mind, we intend to proceed as follows: We will outline briefly the characteristics of Mode-2 knowledge production. Each of our separate presentations deals with three of the characteristics and each ends with a particular perspective on transdisciplinarity. The reason for organizing ourselves this way is that we are from quite different intellectual origins and naturally accentuate different things about transdisciplinarity. We hope this approach will give you a better grasp of the complexity of transdisciplinarity than if we made a single definitional statement.

Some Characteristics of Mode-2 Knowledge Production

Gibbons

Initially, we introduced the idea of Mode-2 in order to create a new way of thinking about science, which is often framed in strictly disciplinary terms. You have heard a lot of it already this morning. Another language is needed to describe what is happening in research. We identified five characteristics that we think are empirically evident. When they all appear together, they are integral or coherent enough to constitute something new – a new form of production of knowledge.

- The first category expresses the fact that contemporary research is being carried out in the context of application. In Mode-2, problems are formulated in dialogue with a large number of interests from the very beginning. The context is set by a process of communication between various stakeholders. That requires great patience. A problem is not formulated outside of the group and, until they come to agreement about what the problem is and how it will be carried out, resources are withheld and research activity delayed. Too much talk this morning was still cast in the framework of science making discoveries and "others," often industry, applying a solution to other contexts. In Mode-2, these two contexts are brought together in a coherent way.
- The second characteristic is that multiple stakeholders bring an essential heterogeneity of skills and expertise to the problem solving process. In Mode-2, we also see the emergence of loose organizational structures, flat hierarchies, and open-ended chains of command. (Rectors, pay attention!) Universities are precisely the opposite type of organizations, which are for the most part highly hierarchical and fixed towards structures. In Mode-2, we find almost the reverse of that.
- The third characteristic of Mode-2 is transdisciplinarity. If we had intended to use "multidisciplinarity" or "pluri-disciplinarity," we would have done so. They are not, after all, very complicated words. We chose "*trans*disciplinarity" for a reason. I will give my reason now, and later Helga will add hers. What we were trying to convey with the notion of transdisciplinarity is that in Mode-2 a forum is generated that provides an distinctive focus for intellectual endeavor. It may be quite different from the traditional disciplinary structure. You and I were brought up in a system where the focus of intellectual endeavor, the source of intellectually challenging problems, arose within disciplines. That may still go on, and long may it continue. However, there are other frameworks of intellectual activity emerging that may not always be reducible to elements of the disciplinary structure. Rather, within the context of application, new lines of intellectual endeavor emerge and develop, so that one set of conversations in the context of application leads to another and another and another. As a result, scholars and researchers may spend a lifetime moving among alternate intellectual frameworks.

Helga, does this fit in with your idea of transdisciplinarity?

Nowotny

I will complement what you have said in a moment but would like to start by saying that in some ways our presentation follows upon what the previous panel has already discussed (see *Chapter 2.1*). I found it quite intriguing that many panelists emphasized the kind of obstacles which often confront those who are engaged in any kind of transdisciplinary effort. Distrust over questions of quality, for example: "What is the quality?," "Are you not lowering the quality of what you are doing?," "Are you not pushing down the average quality?" and so on. Such questions are frequently raised in relation to any kind of inter- or transdisciplinary activities.

*Trans*disciplinarity, to return to different kinds of perspective, has a semantic appeal that differs from what one often calls inter- or multi- or pluri-disciplinarity. And, note that the prefix "trans" is shared with another word, namely the one Michael mentioned already – *trans*gressiveness. If it is true that knowledge is transgressive, then transdisciplinarity does not respect disciplinary boundaries. It goes beyond disciplinary boundaries, and it does not respect institutional boundaries either. In addition, there is a kind of similarity, a kind of convergence or co-evolution, if I may use the terms, between what is happening in the sphere of knowledge production and what we can see going on in the way that societal institutions are developing.

For example, we no longer are in the regime distinguished by the grandiose nation-state so characteristic of modernity, where there was a clearly-structured, highly-differentiated political, economic, and social order with different functions taken up by different sectors of society. What we see today, and again the panel emphasized this, is a tremendous resurgence – especially once we move outside of the European context, which was the home of the nation-state and into the world, say of NGO's – of the importance of various kind of stakeholders in shaping social reality. This is why I think the transgressiveness of knowledge is better captured by the term transdisciplinarity.

Now, there are two more very important criteria of Mode-2 knowledge production, in addition to the ones Michael has mentioned – problem definition, heterogeneity, transdisciplinarity. I will speak briefly about them. They are accountability and quality control.

- Accountability differs from what has been mentioned before, that is individual responsibility. I agree that everyone should have an ethos of individual responsibility. Yet, it is much better if we also have some sense and some sort of institutionalized responsibility. This is what accountability in Mode-2 is all about. Accountability is an informal process. though it also has a formal side. You know to whom you are accountable. There are certain procedures to make things visible that are otherwise invisible. And, it brings in – this is where the link to transdisciplinarity comes in – different groups in society who want to know "What are you doing for us?" Or, to use another term, "What have you done for us lately, you who work in the area of knowledge production?" It is in this sense of accountability to different shareholders, to different users, that the way to understanding how scientific knowledge is being produced opens up. Once there is awareness, which has to penetrate even to curricula and the way we educate future scientists and engineers – once you open this up – then accountability becomes a way to broaden the horizon of those for whom you are producing knowledge.

- Quality control is a very tricky criterion. In *The New Production of Knowledge* we readily admitted that, in the way we described Mode-2 knowledge production. It is, if you want to put it this way, the Achilles heel of transdisciplinarity. What quality control demands in such a setting is not only scientific excellence, which is and remains the basis of producing good new knowledge. It goes further and in ways that are difficult to grasp, because the context varies. One doesn't have a single good criterion as in scientific quality control, where you can always fall back on criteria used in scientific disciplines that allow one to say something is "good physics," "good field biology," or "good botany." You don't have this anymore. Yet, somehow, you have to bring in these additional criteria of quality, of value-added quality. Actually I think we should go beyond "value-added." We should start to speak about "value-integrated." There is a societal value that needs to be integrated into the definition of good science. Since our presentation here deals with the potential of transdisciplinarity the goal is – and the potential exists for is a better outcome, to pick up on something Olav Kübler mentioned this morning – that outcome should be producing better knowledge. It should be better science.

The Novelty of Mode-2

Gibbons

So, there is the nest of characteristics we presented in our first book. It has brought forth some criticism, as you might guess. But, two things need to be said before we press the argument further.

- First, one of the things that characterizes the world in which we live, the transgressivity of knowledge if you like, is that we live in a world where the number of places where recognizably competent research is carried out is now enormous. Just think of the range of laboratories, consultancies, think tanks, research centers, industrial labs, government labs, and universities. At a large number of places, these institutions and organizations now figure in peer review processes. I see this every day. Expertise is being drawn in from a wide range of organizations that are being asked, in my case, about the quality of academic work. This social distribution of knowledge production, the fact that it is spread throughout society, is one of the underlying causes of why intellectual work can go on in so many alternate centers. It is hard for universities in many cases to accept this reality. In Mode-2 knowledge production, the university is a key player. Even so, it is only one player among many because very, very many different strands enter into the context of application, and in the ensuing dialogue, discussion, debate make their contributions.
- Second, and before we move on, we want to clarify one mistake in our book, in our first approach to this matter. Having talked at length about the social distribution of knowledge, the context of accountability, multiple stakeholders, and so forth, a misunderstanding arose: that in some way we were describing an internal development of science. Many of our critics said "Mode-2 already existed in the 19th century" or "Mode-2 only reflects the fact that the NSF is now making many more program grants available." Here, we are stuck back in the language of fixed categories that make a rigid demarcation between society and science. We didn't drive home enough a different point: we are talking about a new mode of knowledge production in a different kind of society.

Nowotny

We have very often been asked "What is new about Mode-2?" or "Has Mode-2 knowledge production not always existed?" To some extent, whenever you invoke a historical precedent, someone always points to its prior occurrence. Yet, we are arguing and are convinced that there is a qualitative jump occurring right now. This jump arises partly because we see how societal institutions are changing. The *ménage à trois* mentioned earlier by Charles Kleiber, had a somewhat different ancestor. Initially, science policy after 1945 was formulated in terms of the state, science and industry. Each knew precisely the role it was supposed to play. Now, though, we see that society is becoming much more differentiated, there are new voices wanting to be heard, and there is a process of democratization going on in societies that does not stop at the door of science. Accountability is also being asked of science, and we see – a term I will come back to later – a kind of "co-evolutionary" process going on in the way that societal institutions develop and how knowledge production develops.

One way of getting back to what Michael was asking me would be to say, "But what is distinctive about Mode-2, if it has always existed?" Here, it is important to say what we do not mean, responding to another voice of criticism that we often hear. We do not say that basic science, fundamental science, is becoming unimportant or that it will go away. We do not make an argument for a sort of untamed capitalism, deregulation, flexibility, or neo-liberalism. We are also sometimes accused of saying that all of the criteria we mentioned are positive: that accountability is positive, and quality control is positive. We stick to our original view, making a further distinction in looking back at what Mode-1 science did. Mode-1 science was highly successful. The way that universities are organized, courses are still offered today, and science has progressed were all based on Mode-1 science. Yet, at the same time, two different views of the practice of science were operative. There was a "front room" view and there was something actually happening in the "back room," on the backstage so to speak.

In the front room – a distinction that goes back to the second half of the 19th century, when our present universities became institutionalized and the take-off with industrialization based upon science began to develop – the distinction between pure and applied science was generated. This distinction formed the basis of how the image of science was projected. It had some interesting consequences. For example, it allowed making a clear separation between fact and fiction, between facts and values, and between knowing what reality was, while avoiding traps of ideological distortion. Wonderful distinctions if you believe in them!

Yet, at the same time – historians of science and technology tell us this very clearly – Madame Curie, to take an example of a female scientist, was not only a great scientist, she was also the founder of the nuclear industry in France. So, there was a kind of engagement between science and society. Many other examples could be cited of the range of engagements going on backstage. The distinction between pure and applied science, which provided an image of science in the front room that was quite different from what was going on in the back room, also had the consequence of letting science off the hook in claiming interest in pure research.

We contend that this interest is hard to identify or to argue why it should be found particularly in science? After all, people are curious. Curiosity is one of the important driving forces of basic science. It's an interest, so stick with it and proclaim it loudly. Yet, realize there are other and many different kinds of interest pressing for attention as well. In laying out this interest, in making it explicit, I think we could rewrite not only the history and perhaps the chronology of science and technology. We also come to a better understanding, a deeper understanding, of what has happened in history, what is actually happening now, and what is needed perhaps even more in the future.

Science and Its Contextualization

Gibbons

To summarize, we have now described a society where not only is science doing the familiar activity of communicating results of its research to society and so transforming it. The 19th-century and early 20th-century model of separate institutional domains is changing to a model characterized by the blurring of boundaries between institutions. In other words, society is moving into a position where it is increasingly able to communicate its wishes, desire, and fears to science. What, then happens, to science and as result of this reverse communication?

First of all, it is a portentous, not a trivial, change. This fact has not been sufficiently grasped. Let me illustrate. We are familiar with the idea of science communicating with society. Much of the debate about public understanding of science presumes that non-scientists are not really *au fait* with the latest developments of science and need to be informed. We expect that form of communication. We are used to descriptions in common sense terms of beautiful discoveries, of developments in instrumentation, and so on. But, once you allow for transgression, and once you allow that institutional boundaries have become fuzzy, you open science to a flow of reverse communications. This is what we mean in our new book by the term "contextualization." Science by its very successes is bringing in its own train, a transformative factor. This process is furthered by a parallel loosening of institutional structures. Furthermore, when society has ways to communicate with science, it cannot expect to stay the same. That's the kind of world that we now occupy.

Nowotny

The term "contextualization" may also sound different chords in different ears, but let me explicate it in a simple though perhaps unexpected way. Contextualization means people. Contextualization means bringing people into knowledge production by asking a simple question: "Where is the place of people in our knowledge?" Taking contextualization seriously means asking that question even in areas of knowledge production that seemingly are far away from people. Of course, you can envision a long line of people forming, especially in a field such as molecular biology, when research has something to do with genetic diseases, or when clinics are involved. Yet, simply asking such a question in every sphere of research will alter the way that knowledge is being produced and make us more aware that a process of contextualization is occurring.

Asking the question of the place of people also implies another thing, an additional dimension. Namely, researchers move not only in the *context of application*. They need to start thinking about the context of *implication*. What are the implications of what we are doing, of formulating problems in this particular way? To stress the importance of the *context of implication* is not a call for new foresight exercises, the kinds of things that have been tried and go on now for a variety of different purposes in different countries. It calls for something much more radical, namely to start asking in scientific laboratories the question, "Where is the place of people in our knowledge?"

People occur, so to speak, in three different variants.

- In one variant are ordinary people whom you encounter every day. They are not statistical averages. They are real people and are becoming more and more part of scientific organizations. NGO's, to recall an obvious example, depend for part of their funding on such people. This is also true for the research that some NGO's carry out. In the medical field, for instance, NGO's depend on funding that comes not only from the state or from industry, but also from voluntary associations and private charities, etc. So, in this variant, people are funders.
- In another variant, there are people who protest against what science produces. Here in Switzerland, we have had the instructive example of the *Genschutz Initiative* not too long ago. This is another way that real people come to influence the way in which scientific development is perceived. (The *Genschutz Initiative* was an amendment to the Constitution proposed by citizens, followed by a referendum to substantially

restrict the field in which genetic research should be allowed. Despite substantial discussion, it was rejected.)
- There is still a third, more subtle variant, that depends on the way in which people are conceptualized in the research process itself. This is obvious in some areas of science, such as environmental science. You cannot do research on problems that have to do with degradation of the natural environment without accounting for human intervention in these processes. You cannot do research in these areas without thinking of policies, or strategies of mitigation and their implementation. So, this is another way that people come in, how you conceptualize them.

Thus, there are many different and, let me emphasize, legitimate ways of conceptualizing people. You can take people abstractly as a statistical aggregate. For example, if you do statistics on problems of urbanization in the Third World, you can ask what these mean for environmental integrity. You are right to look at the statistics of populations living in Mumbai and ask, "How many people are going to live in the next five years in the surroundings of Mumbai? What does it do to the natural environment there?" This is only one way in which people can be conceptualized, certainly not the only way.

There are other questions you might want to address where it would be perhaps better to conceptualize people as active agents – people who have desires, wishes, political preferences; whom you can mobilize to build institutions and to enhance their capacities, and other considerations of this kind. So, there are different legitimate ways of conceptualizing people. We are arguing for being aware of this, of what you are doing, and making it something explicit, so you know what place you give to people in the knowledge you are producing.

Gibbons

So, it is possible to conceptualize people in different ways. An additional pressure favors the emergence of multiple centers of research activity: interests are enormously varied, and the groupings that form the context of application combine these many different interests. One consequence of all this is altering the role of government as a funder of research. It, too, is becoming one player among many in the configuring of resources in particular contexts of application. The institution that funded our latest book has a lot to do with charitable foundations, and it believes there is evidence to show how charitable foundations are beginning to fill some of the interstices of the research agenda in many different countries.

A recent example is the use of medical funds from the Welcome Trust in Britain, which put, I do not remember exactly how many, hundreds of million pounds into biological research. Whatever the level of funding, it is most unusual to find a charitable foundation forming a discrete element in funding of the United Kingdom's national science base. If you want an example of institutional transgression, this one deserves further study. In terms of the language we have been using, the line between the private foundation and the government is so transgressed that the foundation has been able to combine its resources with government.

The Production of Socially Robust Knowledge

Gibbons

A couple of difficult topics need to be addressed. If you accept the view that boundaries between society and science have been well and truly transgressed, what does that mean for the normal way we think of the demarcation between science and society? One of the implications of Mode-2, of course, is that it blurs and makes it harder to say where science ends and society begins. But, the whole epistemology that drives Mode-1 science is based on a very clear separation of science from society. Its epistemology is based on the idea of discrete areas of specialization structured on a particular model of communication. Within each area there are really only two elements:

- first, all research must be able to be written down in a form that can be understood by one's colleagues;
- second, all research must be in a form that eventually can attract a consensus, even if a limited one.

Of course, the propositions up for consideration must be in a language that peers are familiar with. Peers are the ones who make the consensus. Embedded in this model is a notion of reliable knowledge which comprises a whole series of relatively separate decisions about the integrity of a certain set of scientific findings. The limits of integrity are dependent on the limits of the consensus achieved. Indeed, with the growth of specialization, many scientists agree that there isn't any overall consensus among the scientific community, about what the "truth" status of science as a whole is. There is only limited consensus by groups of experts about where consensus lies in their field of expertise.

In a regime where the line between science and society has become well and truly transgressed, how is this epistemology going to guide us? Is it sufficient, for example, if people are drawn into the production of science? What epistemology, then, will guide us?

Somewhat provocatively I ask, "Helga, can a way be found to move beyond merely reliable knowledge?"

Nowotny

Reliable knowledge has served us well, and it is going to stay with us. Without having the internal quality control of a peer group, science cannot be a sustainable enterprise, because there is one clear-cut demarcation criterion to determine "Does it work or doesn't it?" So, reliable knowledge is and will remain the basis of much of science. However, there are many more instances where reliable knowledge is no longer sufficient. Reactions of the public to the results of science are not infrequently contested. There are many controversies now and more to come, because we live in a society where the educational level is high. What is needed is that education strengthen the critical facilities of people. So, we should be glad that we have a highly educated and critical public to engage with in debate.

But, what very often emerges when there is confrontation about a controversy is what many of my scientific colleagues perceive as a refusal of their work. It is a deep narcissistic insult to them, because they work hard and come up with "good" science – good and enthusiastic science as we saw in the beautiful presentation from the US National Science Foundation here today. Yet, somehow the end product is not sufficiently appreciated. In brief, it is refuted or it is contested. This is where our thinking about something more being needed comes in. The answer we give, and it is an answer that needs to be filled out collectively – we do not have a recipe about how to do it – is that we need in addition to reliable knowledge what we call *socially robust knowledge*.

Robustness is a term familiar to engineers, because some of them work on how to make buildings, for instance, more earthquake proof. Robustness is not an absolute concept, but it is not a relative concept either, I would say it is a relational concept. To go back to engineering, robustness depends where your building is located. Is it an earthquake prone zone, or is it an area that so far has been relatively earthquake secure? What kinds of materials are you using? What kind of function will the building have? ... and so forth.

To bring the metaphor into the area of our concern, we cannot predict where the next controversy in the confrontation with the public might arise. We are pretty sure there will be future controversies. Yet, somehow, we have to try to anticipate such controversies and those instances where products of science might be refused or contested. How can this be done? By bringing people, their perspectives and desires into consideration; by thinking more deeply about what people may want and are looking for; by taking their anxieties seriously and communicating with them. There are many local initiatives, initiatives that could bring people into the process of knowledge production and could go a long way toward making knowledge socially more robust.

Re-thinking Science Takes Place in the *Agora*

Gibbons

Now we are coming to the end of this session and I know that you will want some time for questions, but I need to say just two further things.

- First of all, we have been much aided by the work of an American historian of technology, Thomas Hughes, who has written a lot in his recent book, *Rescuing Prometheus*, about the change in ethos amongst engineers, particularly in the area concerned with finding solutions to complex projects such as managing the construction of a building, a highway and tunnels. Hughes comes to the interesting conclusion that over the last 45 years, more and more inputs, including from various pressure groups, have been brought into problem formulation, design and completion of large projects. To us that makes sense. But, what keeps coming back to me when I read and re-read the text, is that this ethos has now acquired a kind of feedback loop. Engineers now realize that you get a better technical solutions if you bring in these views. (See also *David Marks* in *Chapter 6.2.*)

  This shift in the collective ethos of engineers is quite revolutionary. More societal involvement does not mean a better social solution, or a better adapted solution, or a solution that brings social tranquillity to a community. It means a better technical solution. Could not the same conclusion be applied right across the scientific spectrum? Better scientific solutions emerge if there is dialogue with society than if there isn't? Many scientists would, I suspect, have an instinctive bias against this conclusion. Indeed, many would argue that social inclusion will give you weak technical solutions. The evidence presented by Hughes suggests otherwise.

- Finally, if the reverse communication, the contextualization of knowledge, is affecting even the epistemology of science, and, *a fortiori*, what counts as valid knowledge as we have suggested, then you have some idea of the transformation that is involved in the title of our new book, *Rethinking Science*. If we are correct in our analysis, then the real challenge for universities is to take on the production of socially robust knowledge, almost as a crusade. If we are right in predicting that science might expect to experience more or less continuous contestation, one needs universities that will put as much effort into promoting the values, interests and epistemology and modes of behavior of Mode-2 as they put previously into establishing themselves as pre-eminent in Mode-1 science.

Nowotny

I just want to add two more thoughts.

- First of all: better solutions. It is perfectly accepted and considered highly desirable across a wide spectrum of institutions, from industry to policy makers, that innovation, much of what we call the thrust behind innovation, comes from new links between what we call the producers of knowledge and the users of knowledge. Here you have a clear example of where better solutions come about from this interaction. I think this should be the kind of model that is developed, though its basis needs to be extended. We must not just say "Now we have certain categories who are defined as users." The notion of "users" needs to be extended. If knowledge is transgressive, then we need to open up the process to the whole range of reverse communications.
- The second thought goes back to what could be the appropriate structure in which a debate of this kind might take place. We call it, returning to an old Greek term, the *agora*. Creating an appropriate structures of debate requires the management of complexity in public fora, everywhere. You not only have to create a new institutional superstructure or a cantonal structure or a local structure. Debate must occur everywhere. It must be in your mind as much as in social settings, as much as in corporate structures and corporate governance, that it is taken into account as policy making and politics.

3.4 The Responsibility of Science and Scientists

by Richard Ernst, Swiss Federal Institute of Technology (ETH), Zurich, Switzerland and Nobel Prize Winner in Chemistry

Abstract

In recent decades, public interest in research has grown. Many countries have initiated task-oriented research programs. The top-down approach, however, is not sufficient to promote collaborative, goal-oriented research for the sake of society. Scientists must assume greater responsibility. The question of relevance must be discussed more explicitly in scientific circles and a greater public role assumed, including public teaching and speaking out in the media. Responsibility must become a fundamental ethical principle of science.

Introduction

I welcome the opportunity to say a few words about "The Future of Science and Transdisciplinarity." I do not want to repeat how indispensible interdisciplinary or transdisciplinary research is for solving urgent problems. This was evident well before meetings with this subject were organized. But, so far, transdisciplinarity has remained a catchword and still awaits proper implementation.

The various disciplines of science and engineering have been separated not because they are truly independent, but because it is no longer possible for a single individual to master all conceivable knowledge and skills. Division of tasks and of labor is a characteristic of any advanced civilization. A shoemaker does not produce shoes in order to become the most respected member of the shoemakers' guild but is conscious of serving a purpose. The same should be true for science. Indeed, both form indispensible links in the chain of tasks that supports society. Scientists, more than shoemakers, have an inherent desire to become famous. However, true fame is not achievable within the limits of science, only by having a lasting, beneficial effect on society.

Every human being is at heart an explorer, an inventor, a scientist. The instinct of curiosity has been a primary motor of human evolution over the past 300,000 years. Without our child-like drive to explore, we would never have become what we are today. In the course of development, scientists emerged as the experts in curiosity, in innovation, and in understanding. Science, in turn, became the motor of technological progress.

Responsibility in Research

Today, I want to stress a single aspect, the "responsibility" or "accountability" of science and scientists. I speak especially to scientists. During the past few decades, public interest in research has grown significantly. Convinced of the importance of research collaboration, task-oriented governmental research programs have been initiated to focus scientific ventures, such as the EU Framework Programs, and in Switzerland the National Research Programs (NFP) and the Swiss Priority Programs (SPP). This top-down approach is also followed in other countries worldwide. Another trend, especially in former socialistic countries, in developing countries, and curiously in some advanced western states (e.g. Germany and France), is the formation of dedicated government-supported research institutes. The basic idea or fear behind these governmental initiatives is that scientists are still overwhelmingly individualists who have to be told how to make themselves useful to society.

But is the top-down approach truly the most efficient way to promote collaborative, goal-oriented research for the sake of society? We all know how many revolutionary inventions have been the result of serendipity, chance and luck, without intentional planning. Everybody has favored examples. Some recall the accidental discovery of penicillin by Alexander Fleming, carefully observing his infected petri dishes. Others mention the discovery of genes by Gregor Mendel, a breakthrough that would not have happened if his father sent him to a different monastery lacking a herbarium. Or the invention of the Internet, or Post-it notes, or discovery of the fullerenes. There are hundreds of examples of inventions that arose accidentally outside of governmental programs. Perhaps I am unaware, but I do not know of any such invention or discovery that happened inside one of these splendid programs. The conclusion seems obvious: Liberate science from all external constraints. Let scientists pursue their own intentions and satisfy their inborn curiosity, and stop topical research programs. This would most likely not diminish the chance of discoveries and might even save resources by reducing the number of opportunistic profiteers.

But then, we must ask a hard question: Is unrestricted research, free of any conceivable constraints, really necessary? Why do we need further inventions? Why do we even need unrestricted progress? Isn't the world already in a bad state? Do we really have to make it worse or, at best, more complicated? These are questions we have to address honestly. Progress in and of itself is not inherently good.

This is where responsibility comes in. Severe problems torment society, which science is asked to address. Many frightening perils are known, such as HIV/AIDS, Mad Cow Disease, Alzheimers, Parkinsons, and cancer in the health sector; threatening problems in the food domain connected with exhausted soil or lack of safe drinking water; environmental problems that form a special focus of this conference; degrading of historical monuments and documents; and the almost insolvable global social problems that call for an improved and just global economical system.

The demand for responsibility in science requires, necessarily, top-down government programs. At the same time, it calls for transdisciplinary collaboration within a science community that is aware of the urgency of problems in society and is willing to address them, irrespective of personal opportunistic desires, loss of personal financial benefits, or impeding possibilities for greater fame within a narrow speciality guild.

I am convinced that scientists themselves can localize best the problems and questions most relevant for the future of humankind and can judge which ones can be answered with our present means. This belief does not favor applied research. On the contrary, basic questions often have to be answered first, before a practical problem can be addressed. Properly managing research focused on "real" needs requires a lot of foresight. It demands awareness of current trends in society, politics, and industry, and it cannot be accomplished by reading exclusively scientific journals.

Scientists who ask for freedom of research must also assume full responsibility for their ventures. They might be asked, as well, to justify their past and future activities in terms of their utility for society. After all, most scientists are employed and paid by the state and are expected to be in some way useful to the public. At present, the question of relevance is not sufficiently discussed in scientific circles, in lectures, publications, and congresses. Even coffee-break discussions in research groups concentrate more on technical questions, competition, and perhaps science fiction. The link to daily reality on the streets is often not made. Obviously this connection cannot easily be established in all scientifically detailed questions, and it is not necessary that all scientific adventures are socially motivated. Sometimes, pure curiosity is a stronger driving force that leads more rapidly to unexpected solutions to old problems. Still, there should be a red thread of relevance running through all research projects. This thread is a guideline for making decisions at crossroads and, last but not least, for external justification of expensive research efforts. An honest red thread of motivation also helps scientists sleep better (not necessarily at the desk).

Responsibility for Public Teaching

Everybody knows that research and teaching cannot be separated. First, the researcher learns by doing experiments. Research is an essential part of the learning process. Whoever discovers something wants to spread the news, first by publication in respected journals and by presenting lectures, then secondly by teaching students. Science only makes sense when results are communicated. Research activities are, therefore, best associated with universities.

Similar to the ancient master-disciple relationship, all scientists should be involved in teaching. This is the only way to carry out the mission of science. Learning by doing and providing enthusiasm and excitement for research are more important than conveying comprehensive knowledge. Sometimes the disappointing experience of failure, the experience of one's own limits, and proneness to experimental and logical errors are the most educative experiences of all. Often, the most successful researchers are also the most critical and, even more important, the most self-critical.

As indispensible as teaching is within the university, this is not the point I want to stress today. There is an equally important obligation to disseminate knowledge outside of traditional universities. Direct contact of scientists with the general public is beoming more and more indispensible. I am not so much concerned about fund-raising activities to support research but more about preventing misuses of technology by our population. These misuses are due to failing to recognize the consequences of excessive consumption and, too often, thoughtlessness and irresponsibility. Science is expected to remove all obstacles on the consumer's way to the land of milk and honey or to the fool's paradise. Many fellow citizens believe science is capable of solving all problems without causing consumer restrictions. True, there are still advances to come, such as a cure for HIV/AIDS or prediction of earthquakes. But, proper public behavior can already ease damages. Simple HIV/AIDS prevention and earthquake-safe constructions, for instance, might do more than science can ever achieve.

Most major contemporary problems do not call for a fancy scientific solution, requiring additional research. They call for responsible and rational behavior among an educated population. A good, or in fact a bad example, is misuse of fossil energy carriers, preempting resources for future generations and leaving only pollution. We are living on the future accounts of our children. Plenty of technologically feasible solutions to the energy problem are known, but we do not want to pay for them now. The same difficulty applies to the problem of pollution. We know how to avoid pollution but do not implement solutions. We know about malpractices in global agriculture, destruction of fertile ground, and shortages of clean water. Scientifically, we know a lot. We know about problems of monocultures and resulting perturbations of natural equilibrium. To avoid overpopulation, we do not need further scientific research, instead, perhaps, research in sociology and in efficient teaching methodology. Many problems we are facing today do not call for advanced technical solutions, just thoughtful and responsible behavior.

I remain convinced that many malpractices could be prevented by proper in- depth education of the population in science. Those who know causes and can properly analyse consequences might, hopefully, commit fewer ecological crimes and develop a sense of responsibility. Lifelong continuing science education of the general public is of great urgency, if future development of human civilization is to be stable. Thoughtless exploitation of all available technical means will surely lead us to disaster.

Continuing education of the public, then, becomes a primary obligation of scientists (Fig. 1):

We have many means of spreading scientific knowledge in society, such as public lectures, adult-education classes, popular publications, radio, and TV broadcasts. An educational TV channel would be an excellent investment of governmental educational resources. True, the additional obligation will be a significant burden for already overloaded scientists. But, this goal deserves a top place on our list of priorities. Even if these activities do not appear scientists' lists of publications, they will be on our lists of merits.

As always, teaching is a two-way activity. Often, it is the teacher who learns most. The same might be true when facing an unprejudiced audience in a public lecture. Yet, surely research activities will be influenced when we realize we have nothing to say that interests the citizen on the

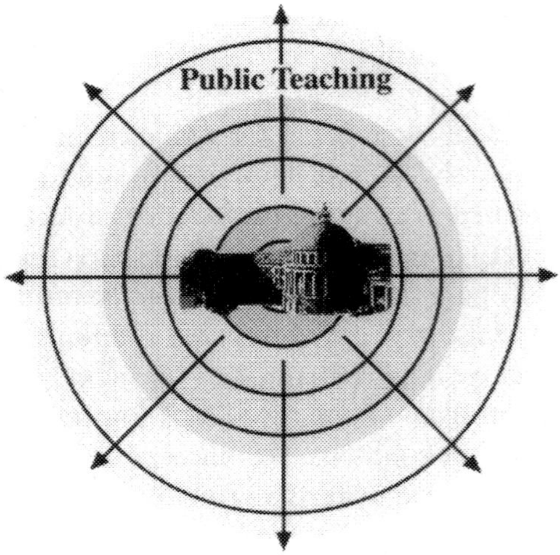

Figure 1
Public Teaching, One of the Primary Obligations of Scientists

street. And, we should never mix science fiction with scientific reality by telling an audience, for example, how important interplanetary missions are for populating planets. That would be dishonest. Diverting people to other planets is not a feasible solution to earth-bound problems.

In general, the present level of education should not be reduced under any circumstances, and requests of right-wing parties to save money at the expense of education are absurd. Life has become so complicated today that everybody has to be properly prepared, even right-wing politicians. Moreover, knowledge should be periodically updated and refreshed. We can ever do too much in this direction. Scientists, claiming to be knowledge holders, cannot simply delegate this enormous task to professional educators. Every scientist has to contribute actively. Public science teaching is one of our primary and most important obligations.

Responsibility for Conceiving and Expressing Conscientious Opinions

Being called upon for more public appearances, scientists should also properly prepare themselves by becoming better informed, conceiving their own well-founded opinions, and developing new and constructive concepts for global development of society. Scientists, especially university professors, are in privileged and independent positions. They have easy access to information. Hence, they have a duty and sometimes even the time to analyse facts and trends. Most important, they occupy almost irredeemable positions and can afford to express their unbiased opinions without immediate negative personal consequences. The same is not true for politicians or for leaders in industry.

Politics is a difficult business where so many compromises are needed in order to reach acceptable decisions. Politics is the art of finding solutions for insolvable problems, something scientists could never achieve. True, there are well respected, honest politicians with straightforward and responsible speech. But, some prefer to express themselves in vague diplomatic ambiguities, and many are trapped in biased and opportunistic party logic and personal struggles. Understandably, they must survive politically. However, there are less encouraging examples of irresponsible and demagogic politicians. Well-known examples include the right wing in Austria, Switzerland, and France. They have lost a lot of public confidence, certainly among the majority of scientists. Even worse are dishonest politicians, such as Helmut Kohl, Roland Koch, and even Bill Clinton. And, a bit nearer to hell are the true criminals, such as Slobodan Milosevic, Franjo Tudjman, Saddam Hussain, Augusto Pinochet, Ferdinand Marcos, Mommar al-Gaddafi, Souharto, Sani Abacha. Who can expect impartial judgments from such politicians?

And what of industrial leaders? Our present global economic system is governed more than ever by the free market concept. Monetary success counts almost exclusively. Optimizing shareholder value is nearly a new religion. It justifies even very short-term thinking. Consequently, employment of personnel has become more dynamic and less secure. Who today wants to express personal opinions that may contradict the money-making but short-sighted strategy of a company, risking their jobs? After all, they have families at home that need to be fed.

Scientists and university teachers are not the only ones who have credentials to express unpleasant facts and recommendations. We must also acknowledge the invaluable contributions of NGO's. For scientists, though, it is not just a right but a paid duty to raise voices of warning. The most important virtue or value in science is honesty and self-criticism. It is perfectly allowable to make errors in science. No real learning occurs without trial and error. Admitting errors is a painful but daily routine for a true scientist, especially when somebody else discovers the errors.

(Better to find and admit them yourself.) Cheating and being dishonest, however, is the worst crime for a scientist. Dishonest people cannot pull themselves out of the mud by their own hair and will be discredited perhaps forever. Without unlimited honesty, science loses its right of existence and can no longer provide a solid foundation for society.

Many scientists recognize what is going wrong in our world but keep quiet for convenience, preferring to work unperturbed in our laboratories or not trusting our effectiveness in public affairs. But if we take our profession seriously, we cannot escape risking the quiet sheltered ivory-tower of our working atmosphere. We must go out into the rain, perhaps even the hail, to proclaim loudly what we think. In this context, it is timely to remember the Catholic monk Giordano Bruno, who was burned to death on February 17, 1600, almost exactly 400 years ago. Bruno expressed his personal and mostly correct opinion on scientific and religious matters, at the time of Copernicus, to the displeasure of the clergy. This event may be put into perspective with current attempts to canonize Pius IX, who declared in 1870 the infallibility of the Pope. As far as I know, no attempt was ever made to canonize or merely to rehabilitate Giordano Bruno. Power and wisdom rarely mate.

Clearly, much is wrong on our globe for which it is worth standing up. I would close by addressing one final concern. Our free market economy, in my view, requires serious revision. The present economical system is not based on morality, on ethics, or on responsibility (Fig. 2).

It is in effect a self-regulating, self-correcting feedback system that operates freely within or at the limits of national and international laws. We turn on the autopilot switch and hope for a safe flight. Everything that is not strictly forbidden is allowed on this journey, and shareholders expect that all conceivable loopholes are taken advantage of to maximize profits. Commercial success is the measure that counts, expressed simply and "objectively" in monetary units.

In truth, we have a great and undisputable role model to follow – nature itself. Nature is also steered by feedback mechanisms, together with a few basic laws of its own, that govern relentlessly over life, death, and evolution. Charles Darwin articulated the laws of evolution, and nature follows them obediently. The stronger one survives, and the weaker one is eliminated. Nature is immoral and unethical, but highly successful. Otherwise, we would not be here in all our glory.

Is this the role model to be followed? Nature had an incredible amount of time, an enormous energy and life were wasted, and the cruelty of nature continues. I would not have liked to be born a dinosaur. I could not have looked with much confidence into the future of my

Figure 2
Free Market Economy, Big Business

species. The same fate was experienced by entire solar systems and possibly their inhabitants. We, or rather our descendents, will see what will happen with our own species.

Anybody acquainted with control theory knows how difficult it is to keep even the simplest feedback systems stable, as soon as a time lag is appreciable. This is why nature has enormous excursions into cruelty. It is also why our free market economical system may run into a dead end. It does not, and probably cannot, take all damaging long-term effects into account. Natural resources are still exploited in an irreversible manner, and social thinking on a global scale seems out of the question. The principle of the stronger surviving and the weaker being eliminated is also valid in free market economy. Adverse side effects on developing countries and on less favored ones in our own countries are, on a short time scale, irrelevant. The gap between the rich and the poor widens and, when devasting consequences finally become apparent, great sums of money have already been made and spent.

Big companies grow and become even bigger, while products become more plentiful and cheaper, and efficiency steadily increases. But, what do we really gain by increasing the efficiency of companies? More luxury at lower costs? Products that are mostly unnecessary and for which first a market has to be created to sell the useless junk that is produced cheaply? Most research investment today goes not into science but into elaborate market research and promotional activities. Not surprisingly, there is a higher demand for clever sales personnel than for uncompromising scientists. I cannot imagine, however, that it is much fun to promote sales of unnecessary products, nor can this way of doing business last very long.

We must express our concern about this kind of economical reasoning. It is necessary to promote, again, responsibility based on ethics and morality in decision making and to abandon purely monetary thinking. Scientists, with their idealism and uncompromising approach, can make a positive contribution, if only by personally resisting the temptations of unnecessary and burdensome luxurious junk. Yet, we should go further, by becoming more outspoken to prevent further misuse of technology and spoiling of the fortunate minority while hurting the unfortunate majority, a pattern clearly visible in developing countries. Simply notice which useless Western products are most frequently advertised in developing countries.

In this context, I urge any scholar to spend, at least once, some time in a research environment in a developing country in Asia, Africa, or South America. This experience can be a true eye opener. By commuting only within Europe or across the Atlantic in the course of our conference tourism, we miss much of the backyard where we deposit our problems. Indeed, many deficiencies in developing countries have their origin within our degrees of latitude. Scientists, who often have more intense contacts with colleagues from those countries, can learn naturally of their experiences and nightmares, and can better articulate the gravity of negative aspects of the western economy. This is one of our responsibilities as scientific globe-trotters.

Experience in a developing country would also reveal the enormous educational need worldwide. There are billions of illiterate people on the globe who have very little chance to adjust to the prevailing economic system. To cite just India: on the one hand computer science is very advanced, and most Swissair software is written there. On the other hand, in the enormous Indian State of Utar Pradesh more than 140 million inhabitants, 86% of the women and 61% of the men, are illiterate. What an incredible task to educate them. Perhaps development aid should go preferentially into education, and scientists could contribute personally by sharing their knowledge with the disfavored.

Following the general trend of free-market economy, numerous politicians have proposed liberating the educational market in Europe and in Switzerland, in hopes of increasing efficiency and of saving money and reducing state taxes. The economic advantages are obvious: we pay only for what we get, and free competition would automatically enhance teaching. But, only those who can pay would get a decent education, and only those subjects would be taught that allow schools to attract plentiful students and thereby stay profitable. An anti-social system would result, favoring a caste-like division within the population that separates those who have from those who lack access to education. The dominance of monetary thinking could start as early as kindergarden or even before, signalling the end of educational responsibility of the state. If a state has any justification for its existence, it is the provision of equal opportunities and initial conditions for all adolescents, necessitating virtually free education.

In our present context, a free market educational system with private universities also implies that unbiased voices of university teachers and scientists would become silent. They could no longer afford to express unpopular truths. Otherwise, their universities would risk losing income and they would lose their jobs. Obviously, quality control is also important in the university sector. But, it has to be implemented with more subtle means than by free-market concepts.

Conclusion: Responsibility as a Fundamental Ethical Principle

To summarize, we live in a period void of ethical concepts. We try to fill the gap where possible with the free-market idea. The market is sup-

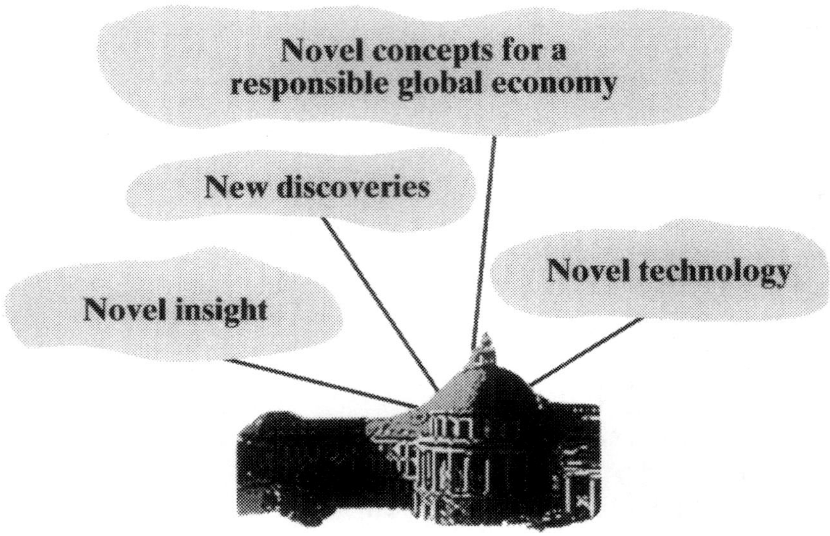

Figure 3
Novel Concepts for a Responsible Global Economy are Demanded

posed to take care of itself. However, it cannot and should not. We must not tolerate cruelties that result from a freely-developing feedback system. We need additional motivations for our actions. They might be derived from insight and understanding, leading to something akin to moral or ethical behavior. We need basic principles, such as a global right of equal opportunity. We need considerate behavior, perhaps even compassion, towards fellow citizens and the rest of nature. We need personal restraints in cravings that unnecessarily exploit the universe. We need goals that do not dominate our environment in any respect or lead to unnecessary waste of any kind. Competition, the basic principle of free market economy, is fine. Yet, it can be deadly when it poisons human relations and destroys nature. Not everything can be approached like sports, where someone always wins and many lose. Other principles must govern human relations.

I am convinced that it is a major task of universities to work towards a renewed foundation of society. In this context, transdisciplinarity is indispensible for fertilizing discourse in the hope of clarifying true goals in science, in economy, and in politics. When universities become places where truly new concepts for society and its future are conceived, the study of the sciences will become even more attractive (Fig. 3).

True, the daily work at the forefront of science is necessarily quite specialized and particular. It easily loses a human touch. But, if it would be combined with the motivation to develop at the same time a renewed foundation for our global society, the work will gain a new inspiring dimension. In this context, the major concept is responsibility, without limits, of everybody and towards everybody. Perhaps responsibility is more important than freedom, and we should work towards replacing the term "free market economy" with "responsible market economy," not simply in our vocabulary but in our actions. We could look, then, with more confidence into the future.

To conclude, let me anticipate what you might think about my sermon: "He would be better off doing something about his alleged problems than just talking." I fully agree. I regret that you have lost another half hour of your active life. If I knew what to do to change the present global trends, I would not give lectures but would turn the crank with all my force and would ask you to help me. Perhaps, somebody in the audience has a practical suggestion where we could start. I am confident there are possibilities. We just have to find them.

3.5 Mobilizing the Intellectual Capital of Universities

by Uwe Schneidewind, Institute for Business Administration, University of Oldenburg, Germany

Abstract

Universities are knowledge-production systems of increasing importance, but their potential for solving societal problems has not been mobilized. Transdisciplinarity is a new way to organize work within universities, informed by concepts of "intellectual capital" and "reflexivity." Transdisciplinarity holds promise for overcoming current deficits of human, structural-organizational, customer, and stakeholder capital, because it integrates different disciplinary, institutional, and community resources.

Introduction

Universities are knowledge-production systems of growing importance in modern societies. But, do they really unfold their potential? Are they organized in an appropriate way to support efficient and effective production of knowledge? These questions are directed at the institutional dimensions of the system of science. Transdisciplinarity can be seen as a new way to organize work within universities and, consequently, is an promising criterion on which to judge these institutions.

Management science has developed the new field of Knowledge Management in order to address questions of this kind in organizations. "Intellectual capital" is a key concept. This chapter will show that it is worthwhile to transfer and to adopt the concept of intellectual capital in universities. Such a transfer reveals many deficits in the science system when it comes to mobilizing intellectual capital. Discussing the institutional future of science means mobilizing its existing potential, a need this chapter addresses by offering ideas for realizing that potential.

Universities as Knowledge Production Organizations and Their Growing Societal Impact

Modern societies are increasingly ruled by unwanted side effects of their differentiated subsystems (such as the economy, politics, law, media, and science). These systems have developed their own running modes or "codes" that enable them to be highly productive. At the same time, differentiation produces imminent side effects in other fields that can't be dealt with anymore in the "codes" of the system (Luhmann 1990). Environmental problems in modern societies are a typical example of this phenomenon. Produced by the overwhelming success (in terms of wealth) of the economic system, these problems cannot be handled in the mode of supply and demand within a price system alone.

In such a context, demand for "reflexivity" increases (Beck, Giddens and Lash 1994; Minsch et al. 1998). Reflexivity means the ability of actors and social systems to reflect on all consequences of their actions and to develop the capability to deal with wanted and unwanted side-effects. The science system plays a crucial role in the search for more reflexivity in modern societies. Being itself a motor of differentiation, by virtue of disciplinary specialization, the science system is at the same time the place where the reflexive power of society is produced and trained in a substantial number of young people. Transdisciplinary science helps to detect, to understand, and to deal with the complex dynamics in and between the different social subsystems (see also *Gibbons and Nowotny, Chapter 3.3*).

At present, this potential is not unfolding in the science system and universities. Reflexive power as a specific, highly developed form of transdisciplinary knowledge has to be produced in universities. A successful outcome will depend on the way that knowledge production is organized in universities. Yet, as this chapter also shows, there are many deficits keeping universities today from meeting the described requirements in an efficient and effective way.

Deficits in Knowledge Organization in Universities – A Knowledge Management Approach

To address institutional deficits of the science system that limit its ability to develop reflexive power, it is worthwhile glancing at the young field of Knowledge Management within management science (Edvinsson and Malone 1997, Nonaka and Takeuchi 1995; Horibe 1999; Probst, Raub and

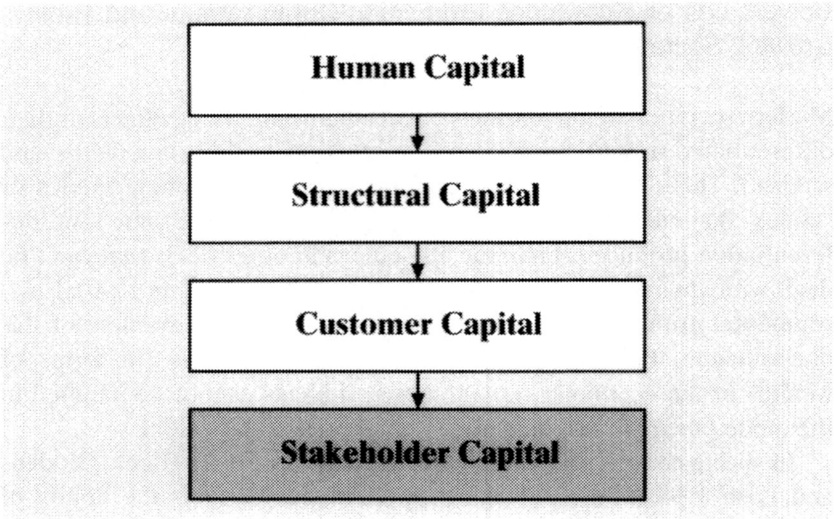

Figure 1
Levels of Intellectual Capital in the Science System (Horibe 1999, Edvinsson and Malone 1997, modified)

Romhard 1999; Willke 1998). Knowledge management deals with the question of how to organize the process of acquisition, use, and growth of knowledge in social systems. Even if the concepts are typically focused on business organizations, they can be transferred with minor adaptations to universities as well (Fig. 1).

"Intellectual capital" (Edvinsson and Malone 1997) is a very helpful concept in this context. It can be seen as a measurement of the knowledge mobilized in an organization. Application of this concept is extremely valuable because of its differentiation of levels of intellectual capital (see Fig. 1). Usually three levels are differentiated:

(1) the intellectual capital of individuals in the organization (human capital)
(2) the structures of an organization that determine how human capital (structural capital) is used
(3) the customer capital that measures the value of customer relationships.

If universities are seen as an integral part of society, a mere focus on "customers" (i.e. students and users of research results in the university

Table 1
Modified Model of Intellectual Capital Levels and its Adaptation to the Science System

Kind of Capital	General Definition	Adaption to Science
Human	Employees' competence, relationship ability and values	Scientists' competence, social and communicational abilities
Structural	Processes and organizational structures to mobilize human capital	Organization of universities and science-networks (incl. funding)
Customer	Present value of customer relationships	Value of students and research partner competences
Stakeholder	Present value of stakeholder relationships	Value of stakeholder impacts

context) is insufficient. A fourth level of intellectual capital has to be added to the model (see Tab. 1).

The different levels are linked in many ways: a good basis for human capital (brilliant researchers and teachers), organized in an appropriate way (structural capital) will build up a huge amount of customer capital (satisfied students, interesting research results for government and business). As a result, a university can attract brilliant staff and invest in its organizational infrastructure (advanced information technology to support the processes).

If we look at the deficits, we can find them on all four levels.

Deficits in Human Development

On the human capital level there are problems of attracting brilliant minds to stay in the science system and not leave for more attractive opportunities in business and private research. Deficits also occur in the social and communicative abilities of scientists, as well as in their personal learning abilities and discipline-crossing capabilities for solving complex problems. The challenge of training these personal abilities more intensely in the future is of crucial importance.

Deficits in Human Development
Attractiveness of the system for outstanding researchers
(Do we have the brightest brains in the system?)
Social and communicative abilities

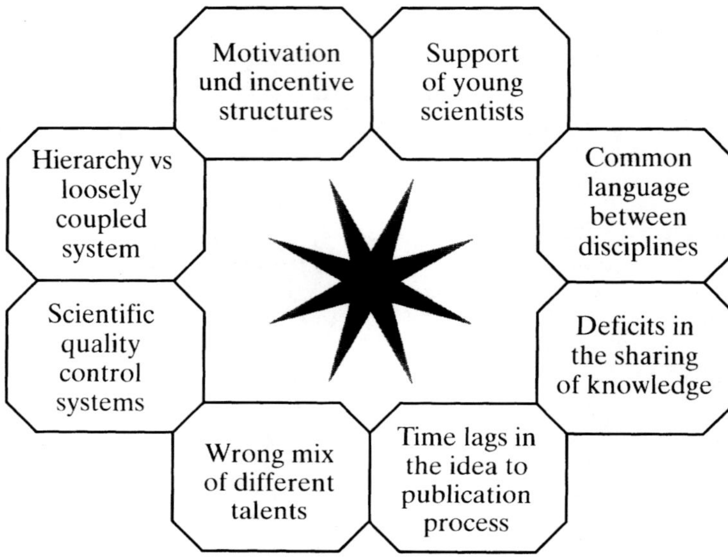

Figure 2
Organizational Deficits (Structures/Processes)

Learning abilities (*Schweizer*)
Understanding the usefulness of complementary skills and approaches
Knowledge of complex problem-solving mechanisms (*Goorhuis*)
Second-order management for non-linear dynamics (*Scheringer and Jaeger*)

Organizational Deficits

Organizational deficits are found in the university system in many places. They are linked to motivational and incentive issues (especially for young scientists), communication and coordination problems, kinds of organizational structures (deep hierarchies instead of flat, loosely coupled systems in business knowledge organizations), and the lack of effective working quality control systems (especially in a new field such as transdisciplinary research) (Fig. 2).

Deficits in Customer and Stakeholder Involvement

The customers of a university are its students and users of scientific knowledge. They are not really involved in the process of knowledge production. Their skills and their motivations are seldom used to enhance the knowledge productivity of the whole system. Frontal teaching (only lecturing) and written exams stifle and fail to support the motivation of the universities' most important customers and its intellectual capital.

Deficits in Customer Involvement
No real use of students' know-how, capacities and motivation (written exam-system)
Form of lectures (frontal, monologue oriented)
Loose connection with users of scientific knowledge (Business, Politics, NGO's)

Deficits in Stakeholder Involvement
No common language between science and society (interface and communication problems)
No common construction of problem contexts (Willke 1998)

In a slightly different way, the same problem exists on the stakeholder level. Stakeholders can influence the knowledge production process in universities in many ways – by bringing in their own knowledge, and by supporting scientists with appropriate societal appraisal or political frameworks which foster scientific research. But, universities very often have not yet developed expertise in involving stakeholders. Many interface and communication problems exist. The current discussion about societal university advisory boards, consisting of different stakeholder groups of the university, is a step in the right direction.

The Institutional Future of Science – Ways to Overcome the Deficits

The four-level concept is also helpful for developing and structuring paths to overcome deficits of intellectual capital within universities. Table 2 gives some suggestions for such paths. Transdisciplinarity is an opportunity in this context, because the disciplines have developed very different levels of proficiency in capital forms. Transdisciplinarity sup-

Table 2
Some Suggestions to Improve Intellectual Capital in the Science Systems on the Four Different Levels

Kind of Capital	Suggestions
Human	Support of social and communicational skills Learn to learn (Transdisciplinarity as a chance)
Structural	New incentive/motivational structures (esp. for young scientists) and funding mechanism Better use of IT-platforms (Information Technology)
Customer	New teaching and research methods involving students (especially in social and transdisciplinary science)
Stakeholder	Involving stakeholders in definition, action and implementation of research

ports interdisciplinary learning not only on a methodological level but also on a knowledge management level.

Conclusions

The science system is the primary knowledge system in society. But, its intellectual potential has not unfolded yet, in comparison to other organizations such as business corporations. The solution is to mobilize its intellectual capital in the system. Transdisciplinary science is a way of raising that treasure. The findings in knowledge management discussed here can help in supporting this process.

4

Interactive Sessions

Note on Interactive Sessions:

Periodic shaded boxes throughout Part 4 contain impressions of the International Conference from eight people, captured by media staff from the Swiss National Science Foundation. These impressions, and an additional statement volunteered by a group of participants, express individuals' points of view in their own voices.

4.1 The Dialogue Sessions

by Julie Thompson Klein

Abstract

The Dialogue Sessions took place at the Swiss Federal Institute of Technology in Zurich the afternoon of February 28th. This traditional format of paper presentations was divided into Fundamentals, Dynamics of Practice, and Illustrative Fields. The summary is based on six themes across one-page reports filed by session Chairs and presenters' texts in Workbook I: Contexts, Methods and Tools, Strategies of Communication and Conflict Management, Programs and Stakeholders, and Theoretical Perspectives.

The Dialogue Sessions

Fundamentals

D01 Theory. Chair, Gertrude Hirsch Hadorn
D02 Implementation. Chair, Julie Thompson Klein
D03 Scientific Quality and Evaluation. Chairs, Paul Burger and Leo Jenni

Dynamics

D04 Complex Systems and Team Processes. Chair, Marie Céline Loibl
D05 Communication and Participation. Chair, Pieter Leroy
D06 Risks and Uncertainties. Chair, Kirsten Hollaender
D07 Intercultural Learning. Chair, Sunita Kapila
D08 Research Programs. Chair, Ruth Kaufmann-Hayoz

Fields

D09 Energy. Chair, Britt Marie Bertilsson
D10 Transport and Engineering. Chair, Hansjürg Mey
D11 Landscape Development. Chair, Michel Roux
D12 Climate and Water. Chair, Jan-Eric Sundgren

Introduction

The purpose of the Dialogue Sessions (D) was to document and to reflect on actual experiences of transdisciplinarity. Participants identified many barriers and impediments (Fig. 1). Yet, they also shared successes and discussed the full range of issues raised by transdisciplinarity. Sessions ranged in size from five to ten presenters, with thirteen to more than forty participants. (All figures contain typical and representative examples, rather than being an exhaustive accounting.)

- conflicts in disciplinary paradigms, methods, and concepts
- incompatible disciplinary, professional, and socio-cultural languages
- ignorance of other disciplines, professions, fields, and sectors
- skepticism about inter- and transdisciplinary work
- lack of systems, collaborative, inter- and transdisciplinary experience
- complexity, different scales, levels and dimensions of analysis

- conflicting constructions of a "problem"
- inadequate information, information overload
- personal attitudes, status conflicts, turf mentality and ethnocentrism
- lack of openness, tolerance, willingness to learn and to collaborate

- professional- and peer-group pressures
- institutional, organizational, and sectoral boundaries
- geographical distance, inadequate funding, time, or infrastructure
- absence of appropriate criteria of evaluation
- inadequate education and training structures
- inadequate rewards, incentives, publication outlets

Figure 1
Barriers and Impediments

Contexts

The Dialogue Sessions spanned a range of problem contexts (Fig. 2). Like the Idea Market and Mutual Learning Sessions, they featured many projects with roots in the Swiss Priority Program Environment. Two sessions were devoted to the specific contexts of energy and transportation. *D09: Energy* dealt with a number of sustainability issues, including green taxes, energy savings in buildings, new technologies such as solar energy and simulation and modeling of electricity demand, and eco-labeling of electricity. *D09* Chair *Britt Marie Bertilsson* identified several prerequisites for success. Adequate time is required to achieve mutual understanding and common goals. A facilitator is also valuable, though presenters advised it not be a stakeholder within a project. An intrinsic motivation is needed as well, underscoring the value of bottom-up collaboration.

- environmental pollution, global climate change
- water systems, atmosphere
- agriculture, forestry, soil
- health care and medicine, social services, aging
- transport, traffic
- local and regional development, urban planning
- landscape development, sustainable development
- industry, banking, professional practice
- energy, electric and nuclear power, waste management
- biotechnology, genetics
- recreation, tourism

Figure 2
Problem Contexts

In *D10: Transport and Engineering*, *Hansjürg Mey* reported, the challenge of selling a large engineering project to the public was a productive framework for discussion. *Ger Wackers and Judith Schueler* compared the Swissmetro project and the Danish bridge over the Great Belt. *Ernst Reinhardt* described efforts to improve fuel efficiency in Swiss road transport, by trying to win drivers' acceptance of efficient driving techniques and new technologies. Several papers dealt with car sharing. In Switzerland, efforts were made to increase the attractiveness of sharing for car owners (*Sylvia Harms*) and the methodology of Strategic Niche Management (*René Kemp*) was tested. A comparable experiment in Berlin was a public-private partnership (*Sassa Franke*), and a project in California demonstrated bottom-up innovation in sustainable transport (*Conrad Wagner*).

Development in the South was a major geographical context. In *D02: Implementation*, *K.V. Raju* presented a holistic model, a Nine-Square

Mandala conceived in a research project on rural farming in India, that recognizes both outer-material and inner non-material spheres of individual and family perceptions of livelihood "security." Development research was a primary focus of *D07: Intercultural Learning*. While it is increasingly transdisciplinary, Chair *Sunita Kapila* noted that few researchers know how to successfully lead teamwork and to integrate findings because of narrow disciplinary training. *D07* was an eclectic group that included an architect, a surgeon, a journalist, linguists, and development students and consultants. Participants from South and North had a remarkable "like-mindedness" on core issues and processes, yet their differences were also a source of intercultural learning. Participatory methodology, *Kapila* emphasized, is fundamental. Both participatory and transdisciplinary methods, though, require more resources and time. So does creation of supportive institutional structures and incentives. The first imperative is adequate resources. In development contexts, donors must also be aware of the working realities of transdisciplinary projects.

Methods and Tools

The Dialogue Sessions documented the great variety of tools and methods that are being used (Fig. 3). There is no uniquely transdisciplinary methodology, and the feasibility of examples in Figures 3 and 4 is dependent on context. Familiar devices, such as surveys, questionnaires, and interviews are common. The use of old tools for new problems, *Felix Rauschmayer* suggested in *D06: Risks and Uncertainties*, opens the possibility of integrating them into a "new" decision tool in a participative process. Because transdisciplinarity is a context-specific, actor-oriented negotiation of knowledge, decision-analytic tools were in abundance, including the Delphi Technique for group communication process, scenario building, and future imaging. Systems analysis is also widely used.

Methodology was a major focus of *D02: Implementation*. In the need-field of nutrition, *Niels Jungbluth* described how a diary study and life-cycle assessment were used to identify constraints and options in consumers' decisions about purchasing ecological food. From the field of social-ecological research, *Thomas Jahn* reported sponsorship schemes, interdisciplinary analysis of problems, and transdisciplinary integration of findings using guiding questions, an integration matrix, clustering into strategic problem areas, and workshops on integration. *Pivot and Mathieu* shared lessons from the interdisciplinary journal *Natures Sciences*

Nothing more than a scientific soap-bubble?

A major and ambitious occasion: The Swiss National Science Foundation, the Swiss Federal Institute of Technology in Zurich, and the electrical engineering company of ABB invited an audience to a presentation of a new research approach, followed by a discussion. The academic baby has been baptised "Transdisciplinarity," having been given birth in the heads of concerned researchers who feel the lack of proper contact with society. They would like in the future to produce socially robust knowledge – a laudable intention.

For media people, it was not an easy conference to attend. It proved to be a difficult job to filter out of this flood of information just those essences that might interest and appeal to a broader public. The two conference document folders alone provided more than 1,000 pages. For the scientific community, admittedly, these events were no doubt very useful. The scientists had the opportunity of thinking deeply once again about their relationship with society, and did so in their erudite, sometimes very elevated scientific jargon. Making only a short visit, I gained the impression that the academic world first needs to reach agreement within itself about what transdisciplinarity is and how the research approach is to be turned into a specific action.

Transdisciplinarity is a long and weary process, and the Secretary of State responsible said that patience and a generous attitude would be necessary. This Zurich conference has taken the initiative and made a start, and we can all hope that this subject will not prove, in our fast-moving times, to be a mere scientific soap-bubble but that it will take on a sturdy shape. Society would have deserved it.

Beat Gerber, Science Editor of the *Tages-Anzeiger*, Zürich

Sociétés (NSS). An integrative outcome in research and editorial work is fostered by establishing a common epistemological field among different partners; translating expertise across disciplinary languages and scales of analysis; clearly delineating a work schedule; paying attention to the composition, structure, and size of a team; and using collective tools such as modeling and collective writing.

Contexts differ, but some approaches are generalizable. In *D01: Theory*, *Eckhard Limpert et al.* illustrated the generic value of log-nor-

- surveys, interviews, questionnaires
- decision analytical tools (Delphi forecasting, scenario building)
- systems analysis
- case study methods, statistical analysis, log-normal distribution
- computer software and networking

- participatory and integrated assessment (Participatory Integrated Environmental Assessment, Participatory Rural Appraisal, Sustainable Development Appraisal)
- impact pathway methodology

- premediation, early clarification, brainstorming,
- cognitive tools of modelling, imaging, and visualization (image analysis, future images, future stories, Sarar method-boîte d'images, backcasting, use of graphs and computer tools, games and other simulations)
- discursive participation methods (mediation, citizen juries, consensus conferences, voting conferences, deliberative polls, focus groups, collaborative agreements,)
- communication tools and strategies (see Fig. 4)

- team coaching, interface teams to bridge disciplines and manage information
- integration matrix, clustering into strategic problem areas
- ecological decision tools, community diagnosis method
- command and control, economic instruments, social marketing instruments
- controlling, oxymoron, Choosing by Advantages
- Strategic Niche Management, project management, flow teams

Figure 3
Methods and Tools

mal distribution for handling empirical data. Transdisciplinary methodologies, *D06 Chair Kirsten Hollaender* observed, aid in both problem orientation and risk assessment. Some participants in *D06* were experts from practice who employ risk analysis as a tool in their everyday work. Multicriteria decision-analysis, consensus conferences, and risk assessment have proved useful as well. There is no one best method, *Hollaender* cautioned. Appropriate methods depend on the problem at stake. They also need to be analyzed and perhaps even changed over time.

In the context of development, in *D07*, analytical tools were also identified. *Kapila* called particular attention to *Hans Hurni*'s description of Sustainable Development Appraisal and Auto-didactic Learning as well as *Desta Mebratu*'s account of using Survival Relevant Information. She also cited an unexpected example: *Thomas Bearth* explored the role of linguistics in development research, posing critical questions through the lens of a discipline that is not common in development research.

Communication and Conflict Management

The quality of comunnication is crucial to a successful outcome. There is no transdisciplinary Esperanto, as *Hollaender and Leroy* observe in *Chapter 6.1*. Inter- or metalanguages are called for, and many established vocabularies, such as systems analysis, are used. New computer software and the Internet also facilitate information access and networking at all levels, from local to international. New technologies, such as Computer-Aided Invention and Common Gateway Interface, aid in building electronic information systems. Oxymoron, a web-based knowledge capitalization and sharing tool, facilitates inter-peer work of students and researchers on specific themes. However, transdisciplinary communication is a collaborative process of translating across disciplines and sectors, creating a new interim or pidgin language for a particular project or program (Klein 1996). Here too, there is no magic formula, but many strategies were identified (Fig. 4).

Gaps in communication occur as the result of obvious differences of language or dialect, different argument styles, and disciplinary/professional and sectoral languages. Thomas Bearth's notion of "communicative sustainability," conceived in the complex multi-lingual context of Africa, is vital to any context, North or South. In development work, the language of target groups has not been viewed traditionally as a resource for solving problems. Yet, oral and written texts and discourses are valuable records of perspectives on nature, society, and history in problem contexts as varied as health HIV/AIDS, ecology (bush fires), agricultural diversification, and democratization of urban and rural populations. Traditional knowledge coded in orally-transmitted language is an asset for mobilizing key groups, who are often women.

- involve lay stakeholders, communities, and political parties early
- engage in pre-learning of disciplinary, professional, and sectoral differences
- define the "problem" collaboratively and establish common goals
- identify pertinent analytical tools and concepts
- disseminate information in language accessible to all parties

- delineate a work schedule with regular (weekly, monthly) meetings of teams

- develop collaboratively team rules, roles, and internal processes
- schedule joint project meetings (twining projects)
- engage in national and international research exchanges
- involve external experts and an independent intermediary

- use communication forums and tools (dialogue learning modules, workshops, discussion techniques, guiding questions, iteration in collective writing)
- use discursive participatory methods (mediation, deliberative polls, focus groups, acceptance dialogues, creation of common vocabulary)
- use dispute resolution methods (use of facilitator or intermediary, consensus conferences, voting conferences, citizen juries)
- use conflict management techniques (conflict transformation, normogramm, project context analysis, evaluation)

- interactive exhibitions, co-publication, study sessions in virtual format
- social compatibility analysis, catalytic encounters

Figure 4
Strategies of Communication and Conflict Management

Because the presence of individuals with different perspectives is a source of conflict, conflict management was a topic across sessions. Heterogeneity and differentiation make friction and tension inevitable. They are not something to be ignored or moved beyond but used creatively. Repression of conflict *D04* Chair *Céline Loibl* warns, is a costly compromise. Special attention must be given to team building and collaborative process. In several sessions, team coaching was advocated as a way of improving the quality of cooperation and communication. There is no guarantee that a particular approach will work. However, transdisciplinary methodology, *Hollaender* added, can help to uncover conflict lines and assess risks, but this does not automatically lead to the solution of conflicts or the reduction of risks.

Workshops have proved valuable in many contexts. In workshops at the journal *Natures Science Sociétés*, collective writing aids in identifying controversies and decodifying the intellectual frameworks of separate disciplines. The Interactive Problem Solving Workshop and European Awareness and Scenario Workshop have also proved effective, and the EVE program (Environmental Valuation in Europe) has used workshops to bring representatives from academia, NGO's, government bodies and international organizations together. They debate a variety of international issues, ranging from human health to the ecosystem (*Carter and Spash*).

Many "private definitions"

In many discussions "in the lobby," the question turned up time and again as to what transdisciplinarity actually is, how it differs from interdisciplinarity, and what specific course of action might be derived from its definition. The organisers of the conference did not attempt to define the term in advance, and left this core question mainly – and perhaps intentionally – to the participants. This gave rise to dozens of "private definitions," and in some cases to scepticism. On the other hand, the conference did raise awareness of the multi-layered complexity of the problem of research work that embraces a number of different specializations, meaning in particular the collision between "soft science" and "hard science" and the difficulties this causes in assessing scientific quality. The organisers sometimes gave the impression that they were making "missionary" efforts to convert "infidels" to belief in transdisciplinarity.

The conference provided an outstandingly good platform for picking up personal contacts with fields of specialization and organizations with which one did not normally have much contact. The result could be described as a "sustainable widening of the horizons".

Prof. Hansjürg Mey, Swiss Federal Commission
for Universities of Applied Sciences, Berne

Programs and Stakeholders

Participants were affiliated with a variety of instituttional structures and forums. By and large, the "home" of inter- and transdisciplinary research and education is not an autonomous institution. Typically, it is a center or institute, program or project, or network. One particular forum was the subject of *D08: Research Programs.* Transdisciplinary research, *D08* Chair *Ruth Kaufmann-Hayoz* stipulated, is not a value in itself. It may be unnecessary or even wrong, depending on the problem at hand. However, it is essential for studying and solving "real world" problems characterized by diverging values and knowledge claims, and a high degree of uncertainty.

The size and scope of programs vary. Larger programs provide favorable institutional structures, while smaller programs and projects appeared to struggle with institutional obstacles. Dutch programs, especially in the presentations of *Jansen* and *van de Kerkhof and Hisschemöller*, appeared to be most advanced in methods and stakeholder involvement in policy research. An Austrian project on Cultural Studies, described by *König and Lutter*, was innovative in developing, or "modernizing," humanities through a transdisciplinary process involving mainly younger scientists.

Kaufmann-Hayoz also reported differences in degrees of stakeholder participation, ranging from weak involvement in the form of advisory groups; to co-funding, consulting on a regular basis, and participation in product or policy development; to the strongest degree of involvement, full responsibility. In addition, the number of stakeholders varied and their degree of involvement over a program's lifecycle. The topic of stakeholders has epistemological dimensions as well. Laypeople and other stakeholders in society contribute lay knowledge, rooted in experience and practice. In this sense, *Hollaender* suggested in the context of the *D06* theme, *risks and uncertainties*, scientists may be seen as lay people who rely on the expert knowledge of practitioners.

In reflecting on the conference goals, *D10* Chair *Hansjürg Mey* raised a question that arose across sessions. *D10* contained good illustrations of "best practices" and ways of promoting and creating favorable institutional structures. Yet, even when presented under its banner, is a project truly "transdisciplinary" if technical facts or real-world experiments are evaluated only by sociological models and tools, or if an analysis is made with no reference to "transdisciplinarity"? To be truly transdisciplinary, a project must make links between different sciences and scientists. Furthermore, *Hollaender* and others stressed, transdisciplinarity entails not just an orientation towards society but cooperation with laypeople and other representatives of society.

Theoretical Perspectives

In *D01: Theory*, Chair *Gertrude Hirsch Hadorn* reported, it was clear that only a small body of common knowledge exists at present. Yet, there are strong bases for theory. *Egon Becker* outlined the different kinds of knowledge that have to be integrated. *Christian Pohl* analyzed reasons for problems in definition and presented a typology of modes of interaction between natural and social sciences in transdisciplinary research. "Holism" is a keyword in theoretical discussions. *Martin Scheringer* defined holism as a methodological orientation in transdisciplinary research. In *D02*, *Shmuel Burmil* explored holistic and organismic aspects of integration and synthesis in the field of sustainable development, contextualized within the larger framework of an Information-Society-Integrated Systems (ISI) model outlined in Workbook I by *Zev Naveh*. More work, Hirsch Hadorn added, needs to be done on understanding the transfer of concepts between different disciplines and between research and practice, including the conditions under which it is fruitful and how concepts are shaped by transfers between different fields of knowledge.

A conceptual understanding of "society" is also important. Society, *Pierre Rossel* explained in *D02*, is a polysemous concept, composed of seven fields that form a transdisciplinary "whole." An individual actor is defined by parameters of social position in a complex and socially-constructed interplay of constraints and possibilities that shape identities, plans, knowledge, and competencies. Within the complex "sphere" of life, *Marc Mogalle* added, complexity reduction is required for focused research on particular problems. Understanding the parameters of the "individual" is no less important. In *D01*, *Peter Schweizer* explained why innovations are not accepted by people, a question that ranges from every-day psychology to scientific analysis. In *D02*, *Joe Brenner* offered a useful exercise for identifying individual attitudes towards transdisciplinarity and their basis, a crucial step for reducing barriers to collaborative work.

Systems theory and analysis were prominent across sessions. The language of systems, *D01* Chair *Hirsch Hadorn* commented, is now in fashion. However, its role as an explanatory concept and appropriate contexts need to be clarified. In *D01*, *Henk Gorhuis* presented a systems approach for second-order problem solving. *D04: Complex Systems and Team Processes* focused directly on the systems concept. Chair *Loibl* distinguished three levels in dealing with complex systems and transformation processes. On the micro-level, research teams must learn to work in inter- and transdisciplinary settings. On a meso-level, the science system is beginning to transform and to create appropriate curricula and institutional surroundings. On the macro level, political transformations in Eastern Europe have had effects on the science system. Urgently needed changes in social-value systems are also called for, capable of organizing global economy in a more sustainable way. Methods were presented for analyzing changing social values by modeling decision-making processes and environmental behavior.

Successfully organized transdisciplinary projects, *Loibl* added, are powerful interventions into local systems. They bridge cultural gaps and make conflicts between different system logics transparent and negotiable. The experience of presenters and participants, she concluded, confirm the value of professional process steering and reflection for optimizing teamwork and problem solving. If project leaders are heavily involved in research and cannot concentrate on project management and process leadership from a thematically independent position, external support should also be engaged.

Learning about a variety of approaches

It was most inspiring to participate in the Dialogue and Mutual Learning sessions of the conference. In my country, Finland, transdisciplinary practice is developing. To improve the dissemination of research results in e.g. climate-policy related research, user participation is encouraged, and there are some good experiences. However, apart form the field of Environmental Impact Analysis, there are as yet no systematic approaches, let alone research on these approaches. People might be working in a transdisciplinary way without being aware of it. Therefore it was interesting to learn about a variety of approaches in research, policy and their evaluation. Some of these can be brought back home to enhance our methods.

In my experience, plenary sessions are more difficult to make into such intensive experiences. In this conference one presentation stood out: Prof. Uwe Schneidewind showed how managing transdisciplinarity is very much a question of managing intellectual capital, including a number of real organizational challenges facing the universities and the science system as a whole.

Dr. Pirkko Kasanen, Työtehoseura-TTS-Institute, Helsinki, Finland

Next Steps

Three areas are critical to long-term promotion of transdisciplinarity – appropriate criteria of quality, educational programs, and ongoing dialogue.

Traditional notions of scientific validity are not eliminated, but they must be enlarged. There is no single, global standard, because transdisciplinary work is context-specific. *Paul Burger and Leo Jenni* reported many models in *D03: Scientific Quality and Evaluation*. The Man-Society-Environment (MGU) program at the University of Basel has defined basic criteria for research. They include integrating stakeholders from outside the university into project teams and joint formulation of a concrete plan for implementing results and monitoring research processes. *Jack B. Spaapen and Frank Wamelink* presented a novel method that incorporates the societal value of transdisciplinary research into evaluation. Groups work toward agreement on relevant contexts, even though different views complicate assessment procedures. *Max Krott* presented "controlling," a social-science based instrument that improves the procedures researchers use to handle the scientific process, producing information to strengthen program management. *Antonietta Di Guilio* described an assessment instrument for inter- and transdisciplinary research, developed in the Interdisciplinary Center for General Ecology at the University of Berne, that distinguishes different dimensions and phases of evaluation.

The second area of need – education – is fundamental to long-term prospects. Members of the MGU program at Basel urged that transdisciplinarity become part of regular university curricula, not just unique programs. *D04* Chair *Loibl* called particular attention to an inter- and transdisciplinary curriculum in Switzerland. Students work together with local actors in long-term projects, while exploring and documenting their methodological experiences. It is never too young to start, *D06* Chair *Kapila* echoed, though relevant materials need to be developed. Relatedly, all departments at university and college levels should make small grants available for transdisciplinary research to insure cross-disciplinary exposure and reference. *D11* Chair *Bertilsson* suggested looking at Management Schools, which are used to working in a transdisciplinary manner. In *D03*, *Nicole Rege Colet* also offered a conceptual framework for evaluating interdisciplinary courses, accompanied by a procedure for evaluating quality in teaching and learning.

The third area is ongoing dialogue. Many particpants in *D07: Intercultural Learning*, *Kapila* reported, chose to continue their discussions the next day, in *M18: Intercultural Learning*. Follow-up regional discussions are needed for tackling the particular challenges of transdisciplinarity in development. The International Development Research Center of Canada is one of several agencies that would be good partners for such an initiative in Africa. Presenters in *D02* started even before the conference, becoming acquainted via an electronic listserv. Their preliminary experience of mutual learning underscores the value of Internet technology as a tool. The desirability of on-going dialogue is a given, *Joseph Brenner* concluded, insuring wider comprehension of the concept of transdisciplinarity and application of pertinent theory and practice. However, he also cautioned, genuine change, will not come of utopian dreams or naive hopes. It requires a sustained commitment to the communication and collaboration begun in Zurich.

4.2 The Mutual Learning Sessions

by Roland W. Scholz

Abstract
Mutual Learning Sessions (MLS) were all-day "laboratories and design factories" held at 18 different sites on February 29th. Individuals from private enterprise, politics, communities, NGO's and academia enacted a basic principle of transdisciplinarity, working collaboratively on designated problem and topics in Industry and Services, Organizing Learning, and Regional Development. Four additional sessions addressed Dynamics and Fundamentals of Transdisciplinary Theory and Practice. Scholz reflects on lessons learned from this experiment, drawing in part on one-page reports submitted by Chairs of the sessions.

Industry and Services

M01 Technology Sharing – ABB Corporate Research Ltd, Baden-Dättwil, Alain Bill
M02 Gene Seeds – Novartis Ltd, Basel, Hans Peter Bernhard
M03 Sustainable Banking – Zürcher Kantonalbank and ETH-UNS, Olaf Weber
M04 Nutrition – Migros, Zurich, Markus Ulrich
M05 Sustainable Tourism – Alpen Region Brienz Meiringen Hasliberg/ BEREG/GEOPROGNOS/SAEFL/SAB, Madeleine Hirsch Jemma and Ueli Stalder
M06 Health Costs and Benefits – Institute Dialog Ethik, Zurich, Ruth Baumann-Hölzle

Organizing Learning

M07 Education – Postgrade Formation in Telecommunication Berne, Federico Flückiger
*M08** Secondary Education – Bildungszentrum Uster, Peter Troxler
M09 Applying Sciences – HTA Burgdorf School of Engineering, Renata Mathys and Jean-Pierre Steger

M10 Medical Knowledge – Horten Center of Knowledge Transfer and University Hospital, Zurich, Johann Steurer

Regional Development

M11 Urban Quality – City of Zurich, Philipp Klaus

M12 Local Agenda 21 – SAGUF and Forum 21 of Illnau-Effretikon, Karin Marti

M13 Landscape Development – Swiss Center for Agriculture Extension and Swiss Federal Institute for Forest, Snow and Landscape Research, Michel Roux

M14 Solid Waste – City of Winterthur and Office of Waste, Water, Energy and Air of Canton Zurich, Wiebke Güldenzoph (ETH-UNS) and Jürg Stünzi

M15 Radioactive Waste – ETH-UNS, Thomas Flüeler

*M16** Contaminated Soil – Municipality of Dornach, Joachim Sell and Stefan Hesske

Dynamics

M17 Participation – Center for Technology Assessment of the Swiss Science Council, Zurich, Danielle Bütschi

M18 Intercultural Learning – Swiss Commission for Research Partnership with Developing Countries, Zurich, Jon-Andri Lys

Fundamentals

M19 Quality Criteria – Collegium Helveticum, Zurich, Martina Weiss

M20 Theory and Implementation – Forum on Transdisciplinary Research, Rapperswil
Philipp W. Balsiger, University of Erlangen-Nuremberg and Julie Thompson Klein, Wayne State University

* cancelled

The Idea of Mutual Learning and Goals of the Sessions

The core idea of transdisciplinarity is joint problem solving among science and society. From this perspective, mutual learning is the basic process of exchange, generation and integration of existing or newly-developing knowledge in different parts of science and society. The term "mutual learning" refers to the concept of mutualism, which is used in life sciences and in societal system analysis. In biology, mutualism denotes a symbiosis and association between two organisms of different species whereby both profit from the relationship. In the social sciences, mutualism is a doctrine of mutual dependence as the condition of individual and social welfare.

The concept of mutual learning in transdisciplinarity originated in one of the ETH-UNS case studies on sustainable regional and urban development (Scholz et al. 1998). In these studies, researchers noticed that orientations toward sustainable development require, indispensably, a process of knowledge exchange between scientists and people living in an area. Soon after the idea of the International Transdisciplinarity Conference was born, it became clear that it should be organized in a different way than traditional meetings. The concept of mutual learning between theory and practice became a substantial part of conceptualizing the conference. The format of "Mutual Learning Sessions" (*M*) became a way of actualizing the concept, providing catalysts for reciprocal knowledge-exchange.

During the early planning phase, inquiries in industry, business, and administration revealed – despite skepticism towards the new vocabulary – that there was a great willingness to organize and to join Mutual Learning Sessions. After a multistage process of approving about 40 potential sessions, 20 sessions were planned and 18 actually occurred. Organizers of the sessions were actively involved in the preparation phase, which included three pre-conference and one post-conference meetings.

Several goals were established for the sessions. The focus was to be on new methods, new solutions and new organizational structures of science and practice. The task was to develop perspectives in and a better understanding of three areas:

- Good practices of problem identification, problem solving and cooperation between academic and non-academic worlds
- Constraints such as mutual trust
- Framework conditions such as platforms for transdisciplinary activities or career tracks for transdisciplinarians both in practice and in science.

The Mutual Learning Sessions themselves represent a platform, establishing an efficient knowledge transfer from science to industry, politics, and other areas. And, vice versa, with exchange being reciprocal. Through mutual learning processes, science may efficiently take part in problem-solving by bringing theory into practice. (For further reflections on "Verwissenschaftlichung der Praxis," epistemology and theory of science, and the concept of mutual learning, see Scholz 1999, *Scholz*).

In addition, conference organizers intended that the sessions focus on current, burning, and relevant questions in society. In order to show how practice values the idea of mutual learning, the sessions were scheduled at locations provided by partner institutions, including private enterprises, the financial sector, authorities, public and private institutes, and non-governmental organizations. The hope was that locations would allow direct problem encounters in a way that mutual learning would be facilitated through joint experiential learning. A joint experience might occur, for instance, through encountering and even participating in waste processing. In general, partner institutions assumed responsibility for preparations.

Topics and Messages of Mutual Learning

A number of core messages and issues emerged from the eighteen sessions. This report focuses on methodological issues, with references to content illuminating the potentials, obstacles or boundaries of transdisciplinarity.

The cluster of sessions on Industry and Services (*M01–M06*) began directly from the subject matter. *M01* and *M02* focused on technology development, with *M01* featuring the China Energy Program (*Bill*) and *M02: Gene-seeds* focusing on genetically-engineered food (from modified crops). Both sessions address large systems – energy and nutrition – that might be fundamentally altered through technology implementation, the focus of which is clearly on improvement. However, in both cases there is a widespread uncertainty and fear about unwanted dynamics and uncontrolled impacts. Moreover, both sessions were prepared by industry. Yet, both fields have complex discourses and have experienced loss of trust among stakeholders in the past. The sessions definitely contributed to a normalization of relationships and discussions.

M01 and *M02* delivered two striking messages. First, industrial key players are aware of the necessity of networking among developers, users, and consumers. Second, there are deficits in the science community's dealing with these issues. The latter message was highlighted by *Dahinden, Bonfadelli and Leonarz*, who revealed that biotechnology up to now has been largely a multidisciplinary activity in natural science that ignores participatory processes such as mediation or consensus conferences. This holds true not only for industry but also for academia.

Nonetheless, the sessions, held in great openness, document the will of representing industries and some scientists to accept the necessity of change.

The financial market is widely considered to be another big system with great vigor but also a limited transparency that may severely hinder different types and conceptions of economic practice. The core of *M03* on *Sustainable Banking* was the question of what defines a sustainable firm (*Weber*). Representatives from banks, consultants, companies, society, issues of the environment and research in various disciplines defined criteria for a sustainable relation between banks and their stakeholders. The major activity in *M03* was creating a business plan for a new company, capable of fulfilling conditions of sustainability from the point of view of all participants and stakeholders, including banks, society and environment. This virtual company can be considered a starting point for devising sustainability criteria, in the form of guidelines for sustainable economic action useful in research and practice. The session was attended by more than 40 participants, with 40% coming from different financial institutions. Working groups produced substantial results that indicate implementation of sustainability in banks has to be a top-down process supported by the leaders. However, the actual working out of results and practice must be conducted at the operational level (line people), not leaders. Bankers and scientists are systems experts of a different kind, and they must work to complement each other.

The topics of *M05* and *M06* – *Sustainable Ski-Tourism* and *Health Costs and Benefits* – had one critical issue in common. Both have been among the most prosperous and wealthy services, but they have also recently come under pressure.

To take *M05* first, a research project on "The Future of Ski-Tourism" (Brander et al. 1995) revealed that issues such as globalization and climate change have resulted in increased aspirations of shareholders (who are not willing to accept return of investments below 15%) and in market change within the new culture of the "fun generation" (*Sauvain*). Both factors have had severe consequences for Swiss winter tourism. At the same time, they open the door for transdisciplinary practice. Regional economy needs both good models of social and environmental systems at local levels and acceptance of action plans by economic and political decision-makers. Papers written in advance of *M05* and the session itself prove that a multi-layered structure is typical in transdisciplinarity for regional studies (*Hirsch Jemma and Stalder*). For that reason, studies of regional development are also good models of transdisciplinary practice (see *Scholz and Marks*, *Chapter 6.2*). Projects in this area

Health costs and benefits: who cares about costs and benefits in health care

Impressions: The transdisciplinarity of the group allowed the very diverse aspects of health-care costs and benefits to be displayed and discussed. Specifically I appreciated the points of view of the lawyer (Prof. M. Baumann) and of public health (Prof. M. Tanner) who enlarged the normally quite narrow boundaries of the "traditional" health-care professionals. Both, the presentations and the work-groups on specific themes displayed the complexity of the topic "costs in health care." They allowed for more in-depth discussions among the participants and for the presentation of specific situations for the illumination of topics under discussion.

I appreciated the rather small group arrangement which allowed for exchange of quite diverse points of view. Furthermore, asking this group to divide into two smaller groups for drafting a position paper yielded intense discussion and allowed for productive work. I found it beneficial that designated members of these smaller groups had prepared provocative statements beforehand and thus fueled the discussion with possible arguments.

I participated in only one day of the conference. Therefore, my understanding of the yield of the whole endeavor is very limited. Being convinced, however, that the future solutions of major societal problems have to be attacked transdisciplinarily, this conference was just a drop on a hot stone, a first step in the right direction. Yet, for my own work and being buried in "monodisciplinary" tasks, no continuation in cross-disciplinary dialogues resulted from this conference.

Dr. Annemarie Kesselring, Institute for Nursing Science,
University of Basel

were often run by geographers, a group of scientists who have recently come under tremendous pressure in some European countries. The members of *M05*, who experienced a joint one-day exercise in the Swiss mountains with representatives from different stakeholder groups, sent an informal message worth citing and thinking about. The message is that scientists will increase efficiency if they reduce complexity and become more "dirty," while allowing incorporation of emotional-creative methods.

As for *M06*, the explosion of costs in the health system and escalation of discussions among pertinent ministries, insurance, doctors' organizations and representatives of client associations has resulted in new key agents entering a well-established field, plus disputes about their significance (*Korczak*). Topics formerly discussed almost exclusively within one profession and a small group of administrators and politicians have become the object of a broad discourse in society. In the face of unequal opportunities to benefit personally from the health system within and between countries, fundamental questions of social justice arise, as well as the pragmatic issue of conflicting interests of different parties and the challenge of arriving at a consensus about what role they should play. These issues are quite comprehensible, if we take Switzerland as an example. It has one of the best managed-care systems and spends "significantly more than 10 percent of the social income to the health system" (Britt 2000). Yet, fundamental questions require answering. Which system of medical care is desirable? How much are we willing and able to spend? Which indicators are appropriate for which question? The informal message *M06* sent was highlighted by *Baumann-Hölzle*. Being transgressive is a necessity, in the sense that for the future of a socially-acceptable and just health system, some current principles of law and order should be questioned and critically disputed. This dispute, she added, will go far beyond ethics.

In session *M11* on *Urban Quality*, participants from European, African and Asian cities demonstrated their experiences in transdisciplinarity. It was held in Zurich West, one of the former industrial areas of Zurich that is now undergoing rapid change. Most participants had some experience with participation, and a series of participatory planning projects has already been conducted on the Zurich West site. Nevertheless, two questions need to be addressed: how is participation best organized, and which role should generic bottom-up processes play? Various examples were discussed and presented, including the KraftWerk1 project, an alternative apartment cooperative that attempts to reestablish the vicinity of working and living in modern urban contexts. Indicative of the commitment to community, lunch was hosted by youngsters of the city's refugee organization in an abandoned factory.

Session *M12* was held at Illnau-Effretikon, a small town close to Zurich. It began with a presentation by representatives from Local Agenda 21 of the town. In the ensuing discussion, the role of science in these processes was considered. Participants concluded that science

Inspiration gained from deficiencies

I was very surprised to hear in some opening speeches that it was the first international conference on transdisciplinarity, as this must have been a misunderstanding.

The plenary sessions sometimes reminded me how useless it is to invite honorary "big names" just for the sake of it - they never bother to prepare for the subject of the conference. That is especially true of the sponsors and global NGO's. There were, however, very well-selected speakers who seemed to know what they were speaking about, such as the tandems of Nowotny-Gibbons, Scholz-Marks and – to some extent – Smoliner-Häberli.

The mutual learning session (*M11*) was a grave misunderstanding throughout: it was arranged to discuss local problems, very interesting as such but having nothing to do with transdisciplinarity as presented. Foreign participants did not know anything about the situation and felt as if being forced to observe some local fights. Furthermore, I strongly suspect that 97.5% of participants did not have the slightest idea what the definition of the subject was.

At the dialogue session on the implementation of transdisciplinarity (*D02*) which was very interesting and open, I met most of the people to whom I kept talking throughout the conference. Several of us are now considering to meet again this fall to discuss some of those issues further on. Some of us felt there is something to be added to the conclusions of the conference and have put together a statement that may be found at http://transdisciplinarity.net.

Nevertheless, I am very grateful to have had a chance to take part in the Conference. In a way, it changed my life, as communication with other participants confirmed my beliefs and way of thinking in certain issues of transdisciplinarity; it greatly enriched my perspective and provided me with a chance to meet some people whom I admire and love to exchange knowledge with after the conference – even if the cause for exchange is something we missed at the event itself.

Paulius Kulikauskas, Byfornyelse Denmark, Copenhagen

could support Local Agenda 21 processes in a procedural and a substantial way. Methods of integrating knowledge were formulated in structuring, planning and evaluating, as well as appropriate ways of communicating in local development projects. Two related assumptions are inherent in the working definition of transdisciplinary research in this session: local actors should be considered experts and communication should be in a simple, comprehensible language. This new vision of science should be accompanied by a redefinition of science itself.

M13 on *Landscape Development* was described as a prototype of transdisciplinary processes. This description refers not only to evaluation and planning of landscape. The forming and realization of landscape has long been affected by a multitude of different agents, including landowners, media and politicians, as well as agricultural, environmental and construction associations. However, the degradation of biological and landscape diversity indicate the results have not always been positive. Various models of landscape development have been proposed in the past. This Mutual Learning Session was distinct in providing motivation for developing strategies to incorporate users of the landscape in discourses pertaining to both quality management and the costs of landscape use (*Heeb and Roux*). In regard to international practice (Woodhill and Röling 1998), the session also discussed strategies for developing awareness and responsibility towards landscape as a common good.

Waste management is a well-established industry today. Conference organizers placed *M14* and *M15* on *Solid Waste* and *Nuclear Waste* under the landscape section for a particular reason. Disposal of waste – an unwanted by-product of manufacturing, biological, chemical and nuclear processes – is spatially relevant. Both the toxicity of waste and landscape changes (waste causes wastes) are matters of allocation and, therefore, societal dispute. Disposal of nuclear waste can be seen as a negative prototype of societal problem solving, producing deadlocks in discourses and decision-making. Many mediation and arbitration procedures were attempted in the past and most failed.

The sensitivities of current susceptibility of waste issues were apparent in preparations for *M15*. Although all of the invited institutions participated and some provided financial support, none was inclined to play the role of host. Each suspected the session would be perceived as opinionated if it were run under the direction of a key agent. This is why ETH-UNS and one of its external Ph.D. students took the lead, with key players from Forum Vera, Gewaltfreie Aktion Kaiseraugst, Greenpeace, and NAGRA playing cooperative support roles. *Thomas Flüeler* remarked on this fact, and the location of the session was indicative as well, taking place on the site where the first Swiss reactor was chosen to be built 40 years earlier, the ETH in central Zurich. The session highlighted the socio-technical nature of "radwaste" management. The special challenge here is how to integrate an obviously long-term horizon into the decision and implementation process and to balance it with protection and other sustainability issues. There were successes but no basic breakthroughs in the current deadlock. All of the participants' views received equal attention and differing perspectives were aired, most likely due to moderation by a professional, independent facilitator. The most challenging problem, though, remains – how to integrate the main principles of sustainability (passive safety and control mechanisms) into an open disposal system characterized by slow degradation. As indicated in a booklet on *M15* (*Flüeler*), the lessons learned from this session can be summarized and developed elsewhere.

New information technologies also provide a potential for organizing mutual learning and for communication about critical issues. This was the focus of three MLS's, especially *M07* and *M10*. Furthermore, the focal point of *M04* on *Nutrition* was facilitating transdisciplinary science activity with the help of new technologies. The general target of *M04* was to reveal and to communicate hidden structures in and impacts of actions in the need-field of nutrition, using the tool of simulation and gaming. A group of 20 participants had a joint experience of the simulation game, which presented all of the environmental impacts caused by production of one menu (*Ulrich and Jungbluth*). The game fostered discussions and evaluations that factored in emotions and affective responses. Participants took the roles of farmers, food industry, food stores, and consumers, thereby experiencing at least in part constraints on these diverse roles.

M07 on *Education* made it clear that education science, in particular didactics, is a field where the idea of transdisciplinarity has been celebrated for a long time without explicit notice. This holds particularly true for virtual learning systems in which computer developers, teachers from different subject areas, and students are involved in product development. Transdisciplinarity and communication of scientific data and information are also being introduced into the theory of knowledge. This development is acknowledged by the emergence of an intriguing question – How can we cope with the tremendous amount of information produced by academic institutions?

In *M10*, another example of transdisciplinary science production was revealed – allowing the patient to profit from medical interventions. Here, the problem of communication between producer and user of knowledge arises. This problem has been discussed for a long time in science but has never been sufficiently solved for practical purposes. Because this problem is of highest societal and economic significance (see above *M05*) a follow-up project under the guidance of *Johann Steurer* will refer to these questions (<www.evimed.ch>). *M09* on *Applying Science* also explored a topic *Scholz and Marks* address in *Chapter 6.2*: new conceptions of the engineer are indicative of the need for transdisciplinarity to become the object of professional training in higher education.

Participation is also a salient feature of many transdisciplinary projects. The question of how lay expertise can be introduced into the debate on technological and scientific developments was discussed in a large group of about 50 scientists who attended *M17 Participation*. In principle, laypersons or stakeholders should also be interested in participatory processes in which they are expected to play a role. However, no laypeople or stakeholders joined this session. This fact underscores how difficult it will be to discuss issues of transdisciplinarity if no substantial problems of public concern are addressed.

North-South research partnerships have applied transdisciplinary approaches such as joint implementation for more than 20 years. Given this experience, it is no surprise that *M18* on *Intercultural Learning* provided a series of intriguing insights, recommendations and conclusions. Intercultural transdisciplinary exchange requires personal courage, interpersonal trust and time. These constraints contradict trends in science production. As a result, transdisciplinarians run a risk in their push for crossing boundaries. In principle, intercultural exchange and collaboration have the potential to enhance the richness of individual cultures. However, there is also an opposite potential. Transdisciplinary efforts may end up with an unidentifiable mix of different types of knowledge and statements. Thus, participating researchers need their own, firm cultural and disciplinary backgrounds. This recommendation echoes an important lesson learned from interdisciplinary research – interfaces need faces. Finally, one must be aware that North-South partnerships cannot be carried out on an unequal footing: unequal distribution of power among the partners involved remains a concern.

The issues at the heart of *M19: Quality Criteria* and *M20: Theory and Implementation* were discussed in various Dialogue Sessions, panels, and contributions throughout the conference. Both sessions were designed for internal academic discussions. The dialogue between arts and science was targeted in *M19*. An intriguing discussion ensued on the necessities and cultures of doubt. A core question was whether doubt is a privilege and what are the differences between doubt in arts and the productive hypothesis in science. Both concepts have the same function, participants suspected, though they arrived at no final answer. E. Ruhnau of Munich stated that there is a tendency of Western philosophy to create words and then to think that a term has concrete substance already. This remark may have same significance for transdisciplinarity. In *M20*, the need for a platform of knowledge about transdisciplinary theory and method was the major focus. Improving dialogue and practice were continuing threads in the session, which contributed to building the platform that is needed.

Additional insights come from the cancellation of two sessions. One, on Contaminated Soil, did not materialize because of a critical stage of development with privacy concerns among a number of key agents. These concerns conflicted with the original plan for open discussion of the current situation at a particular site. The other session, on Secondary Education, was initiated by a small private company that, in the end, lacked sufficient financial resources for hosting.

Conclusion: General Mission

All in all, the Mutual Learning Sessions constituted a successful format, engaging nearly 800 people, among them 300 people prom practice in intense discussions, disputations, and learning from each other. Some follow-up initiatives were even started in the sessions on Sustainable Banking, Radioactive Waste Management, and Health Costs and Benefits, and presumably other sessions as well.

In addition, mutual learning among members from science and society took place in many sessions. It is extraordinarily difficult to break through traditional forms of communication in conferences. However, the sessions showed this is possible, at least if one starts from the substance matter of a real-world problem of public concern. Lots of ideas and recommendations were conceived for the development and practice of transdisciplinarity, as this report testifies. What seems most important is that some of the sessions initiated fertile discussions and follow-up processes, within academia, within practice and – above all – as joint ventures proposed by transdisciplinarity.

Career opportunities needed

The Mutual Learning Session on "Knowledge Transfer in Medical Care" provided an opportunity for picking up interesting contacts.

Basically, the workshop was organized on transdisciplinary lines but the representatives of "Society" had difficulty in making their contributions. Technical discussions did not really start until the "pecking order" had been established within the group.

All in all, the discussions were lively and interesting, but all the same I find it hard to believe that any specific project will emerge from them because all the participants are already engaged in other matters. The requirement that would have to be met before any transdisciplinary research projects can be started would be the creation of an institution which methodically supported and coordinated these research projects and motivated young researchers by giving them career opportunities.

Being a layman in transdisciplinary matters, I had expected to be given a comprehensible definition of transdisciplinarity. Also, this conference concentrated too strongly, in my view, on problems relating to the environment. All in all, this event was a stimulus for thoughts about possible future research models.

PD Dr. Johann Steurer, Horten Center for Practical Research and Knowledge Transfer, University Hospital Zurich

4.3 The Idea Market

by Walter Grossenbacher-Mansuy

Abstract
The Idea Market was an open forum for networking and exchanging new ideas, tools and experiences. Located along the North and South wings of the ETH, it featured posters, video animation, computer demonstrations, graphic models, a co-presentation with students, and a simulation game. Presentations were divided into (1) Theory & Methods and (2) Examples. Grossenbacher-Mansuy summarizes four themes: Network Initiatives, The Interface between Science and Society, Tools, and Examples.

Goals and Description of The Idea Market

The Idea Market (I) had two main goals:

- to provide an open forum for discussing and exchanging new ideas, tools and experiences in transdisciplinary research
- to intensify the networking process among the growing international transdisciplinarity community.

The actual Market was divided into two sections: *Theory and Methods* (*I01*) and *Examples* (*I02*). Contributions were arrayed along the North and South wings of the main building of the ETH. Besides traditional posters, there was a rich variety of exhibitions, including video animation, computer and on-line demonstrations, model demonstrations, a co-presentation with students, even a simulation game presentation.

The following overview summarizes the content of the market around four themes: network initiatives, the interface between science and society, tools, and examples of research processes.

Network Initiatives

Two network initiatives were presented:

University Institute Kurt Bösch (IKUB) database

This newly developed database contains information on institutions and people, research (projects, programs, and fieldwork), education (teaching and courses), and publications in the field of inter- and transdisciplinarity at Swiss universities. The goal is to facilitate exchange of information among experts, novices and other interested individuals on the topics of research and teaching. Both recent publications and classical texts are included. Further information on the institute, its personnel, research and educational activities, and pertinent publications is available in Workbook I (*Perrig-Chiello, Perren and Eckstein*) and in electronic format at <http://www.ikb.vsnet.ch/>.

SAGUF-network for transdisciplinary research

To encourage further development of transdisciplinary research, the Swiss Academic Society of Environmental Research and Ecology (SAGUF) launched the SAGUF-network (SAGUFNET). It is situated at the MGU (Man-Society-Environment) coordination center at the University of Basel. The network involves both researchers, especially younger ones, and their non-academic partners in transdisciplinary projects in Switzerland. Researchers from other countries are welcome as well. Further information on SAGUF is available in Workbook I (*Förster and Hirsch Hadorn*) and in electronic format at <http://www.unibas.ch/mgu/sagufnet/> (see page 21).

In addition to the two initiatives that appeared in the Market, the Centre International de Recherches et Etudes Transdisciplinaires (CIRET) maintains a global Internet network run by Basarab Nicolescu of the University of Paris, at <http://perso.club-internet.fr/nicol/ciret/>.

Manifesto: a broader view of transdisciplinarity

Contemplating the International Transdisciplinary Conference held in Zurich, February 27–March 1, 2000, the undersigned have decided to call the attention of the participants at the Conference, and other audiences, to our firm belief in the need to place the human being, in his/her different levels, at the center of concerns of Transdisciplinarity in science and society.

We, the undersigned, further emphasize that:

(i) the fundamental principles of transdisciplinarity encompass both the inner and the outer development of the individual, such as: competence in the field of the true vocation of the individual, ethics: commitment, responsibility and accountability, and spirituality in the open sense, as laid out in the Charter of Transdisciplinarity adopted at the First World Congress of Transdisciplinarity in Arrabida, Portugal, November 2–7, 1994;

(ii) fundamental statements on transdisciplinary education are: to open education towards an integral education of the human being which transmits the quest for meaning; to make the University evolve towards a study of the Universal in the context of an unprecedented acceleration of fragmentary knowledge; to revalue the role of deeply rooted intuitions, of the imaginary, of sensitivity, and of the body in the transmission of knowledge, as stated in the conclusion of the International Congress "What University for Tomorrow? Towards a Transdisciplinary Evolution of the University" in Locarno in 1997.

The full statement is available at <http://transdisciplinarity.net>.

Joseph E. Brenner, Les Diablerets, Switzerland;
Paulius Kulikauskas, Byfornyelse Danmark, Denmark and Lithuania;
Maria F. de Mello, Transdisciplinary Educational Center Escola do Futuro, University of São Paulo, Brazil;
K.V. Raju, Institute of Rural Management, Anand, India;
Americo Sommerman, CETRANS, Escola do Futuro, University of São Paulo, Brazil;
Nils-Goran Sundin, Collegium Europaeum, Stockholm, Sweden

Interface Between Science and Society

The Relationship of Science and Policy

The relationship of science and policy is one of necessity. The way a policy problem is structured, *In't Veld and de Wit* emphasized, determines the role of science in policy making. They recommend that scientists begin a dialogue with policy by conducting an inventory of different problem perceptions, setting the ground for confronting differences and ultimately achieving integration. Strategic use of knowledge is common, making the struggle for knowledge equal to the struggle for power. This reality of transdisciplinary work should always be kept in mind.

Language was also a focus. In studying the role of different actors' perspectives, *Moser et al.* analyzed how language functions in processes of knowledge management. Information and interaction are two key concepts. For a successful exchange between scientists and non-scientists, developing a common language is crucial. Two other contributions offered new concepts for transdisciplinary metalanguages. *Cslovjecsek* presented a sound-oriented starting point, moving beyond more customary visual orientations. *Lunca* proposed a transdisciplinary metalanguage on the level of crossing and unifying a number of disciplinary languages. In addition, *Jaeggi* affirmed the need for a common language if the university, the economy, and policy want to profit from each other. The university must also be capable of translating academic knowledge in a manner that is accessible to all participants.

Institutional structures are equally important. *Gandolfi* proposed creation of a European Academy for politicians, enabling politicians from all countries to meet, to exchange experiences, to learn how to think in a more transdisciplinary and systemic way, and to gain knowledge about current political debates on scientific and technological issues. Systems thinking and complexity science, in particular, can improve their decision making.

Because information on nature and the environment often comes too late for inclusion in policy making, *In't Veld and de Wit* urged joint formulation of research needs. The values of science and policy should be explicitly clarified in the beginning and negotiated throughout the research process. This negotiation, *Perrig-Chiello, Höpflinger and Perren* stressed, fosters both research and problem-solving among all actors. Readers interested in the topic of participatory methods will also find Mutual Learning Session *M17: Participation* and Dialogue Session *D05: Communication and Participation* to be helpful.

Because North-South research partnerships are especially sensitive to the dynamics of intercultural communication and participatory research methods, it is no surprise that outstanding examples of transdisciplinary research processes stemmed from this international perspective. See particularly the Idea Market contributions of *Flury, Geslin and Salembier, Jenny and Baumann*, and *Robledo and Sell* as well as the two sessions on *Intercultural Learning: M18* and *D07.*

Consequences for Science and Education

The Idea Market exhibited a wide variety of experiences in methodology and project management, topics also addressed in *M20: Theory and Implementation* and *D01: Theory*. In Switzerland, the Swiss Priority Program Environment (SPPE) is a milestone for transdisciplinary research. The market featured a number of projects that have roots in the SPPE, which was supported by the Swiss National Science Foundation from 1992 to 2001. The examples include *Baumgartner, Di Giulio et al., Flury, Hesske and Frischknecht-Tobler, Künzli et al., Martinez and Schreier, Ulrich and Borner*, and *Winistörfer et al.*

The transdisciplinary approach to knowledge production has consequences for education as well, in traditional universities, the curriculum, and teaching methods. The Idea Market included reforms and innovative teaching approaches in the fields of environmental sciences, general ecology, civil engineering, and industrial ecology (*Frischknecht et al.; Künzli et al.; Lima et al.; Keitsch and Wigum*; and *Stuhler*).

Additionally, *Caetano et al.* illustrated how transdisciplinarity can foster a common conceptual framework and new tools that scholars of different disciplines can use. They stress the necessity of integrating time (history) in social sciences and the plurality or multidimensionality of a given problem. Critical dialogue across disciplines implies redefinition of disciplinary organization and methodological concerns. Additionally disciplines have to be aware of the importance of new international and organizational dimensions.

Transdisciplinary environmental education is an added tool with several benefits. It raises awareness; establishes communication among science, industry and society; and initiates participation in problem solving and decision making processes. The university is not the only level affected. *Hesske and Frischknecht-Tobler* co-presented with scholars attending the Teacher Training College in Sargans, Switzerland a case study focused on the real-world problem of soil pollution.

Transdisciplinary Tools

Decision making is a successful field of application for transdisciplinary approaches. *Craye* cautioned, however, that scientific expertise based on transdisciplinary knowledge provides information for the debate. It does

How does one reach other people?

My impression was positive. I believe that the conference gave the participants new impetus, particularly with regard to the involvement of socio-economic factors in scientifically oriented research.

Two problems still remain: the definition of transdisciplinarity lacks focus, and the participants at the conference were those who are aware of the problem anyway. So the question arises: How does one reach other people?

Dr. Rainer Gerold, Director of Life Sciences and Quality of Life, Commission of the European Communities, Brussels, Belgium

not decide the outcome of a debate. This condition of practice underscores the importance of a two-way dialogue between scientists and policy makers as well as other practitioners from industry, organizations or administration. Clarifying separate viewpoints is a major step towards informed decision making.

Van Veen and Ouboter displayed two different tools: interactive policy renewal and market-driven knowledge transfer. In both cases, transdisciplinarity has proved to be a success factor. Public private partnership and platforms are useful management instruments to stimulate interactive cooperation processes. *Winistörfer et al.* presented the Social Compatibility Analysis, a suitable tool for use in participatory processes such as acceptance dialogues. By visualizing different evaluations and viewpoints of various interest groups, a common basis for discussion and solution finding can be established.

Simulation games are also a successful tool for communicating research findings to concerned actors and for raising sensitivity to multiple perspectives. *Ulrich and Borner* demonstrated their use in the Idea Market, and it was discussed in *M04: Nutrition*. *Villa et al.* presented additional thinking and learning tools using new technologies in the pre-conference Workbook I, available also at <http://www.lnh.unil.ch/>. It contains other pertinent contributions by *Canali, Hoffmann, Limpert et al.*, and *Schweizer*.

Last but not least *Gökalp* underlines the experience of a crucial tool to foster transdisciplinary research: the geographical proximity of all actors. Despite the Internet, face-to-face contacts remain important.

Examples of Transdisciplinary Research

Several members of the Idea Market demonstrated concrete experiences with transdisciplinary research processes. The spheres varied:

Ecosphere:	*Kallis and Coccossis* and *Robledo and Sell*
Sociosphere:	*Bährer, Flury, Geslin and Salembier, Hesske and Frischknecht-Tobler, Petersen and Schaltegger*, and *Vahtar*
Technosphere:	*Adey et al., Hermanns Stengele and Schenker, Jeffrey et al., Jenny and Baumann, Kostecki,* and *Moritsuska*

By integrating all relevant actors in a problem solving process, the possibility of finding solutions – of generating what Helga Nowotny calls "socially robust" knowledge – is increased, though is not guaranteed. Corresponding criteria such as "consensus," "achievability" and "resilience" increase significantly, a major finding of *Jeffrey et al.* The idea of "partnership" between scientists and practitioners in the research process corresponds closely to the prerequisites of sustainable development. Here too, it is no surprise that sustainability research was a major theme in the Idea Market. Illustrations include *Adi, Di Giulio et al., Flury, Hesske and Frischknecht-Tobler, Künzli et al., Martinez and Schreier, Petersen and Schaltegger, Stuhler, Ulrich and Borner*, and *Winistörfer et al.*

Conclusion

The goals of the Idea Market were achieved. The networking process was stimulated, and ideas were exchanged. Three conclusions emerge from the experience.

- 1st: Transdisciplinarity is a concept that is interpreted in a very broad sense. The wide variety of approaches includes the fields of complex systems research, participatory research, and engineering, to name just a few notable examples.
- 2nd: An even greater variety of fields exist in which a transdisciplinary approach is applied. The core idea – different academic disciplines joining together and jointly solving a real-world problem together with practitioners – is not bound to any single thematic field.
- 3rd: In such a heterogeneous situation, network initiatives for both researchers and practitioners are needed.

Conditions for happiness

As the result of the extreme specialization of the various branches of science and all the other fields of activity, as well as the pluralism of values, our present-day society is heavily fragmented. Therefore, individuals have a wider range of personal choices and more opportunities for their own development. But they also need sufficient intellectual, emotional, and social abilities to exploit this potential and to move around within a complex social environment. This requires the ability to see the big picture and to be able to construct one's own world by relating some of the fragments to each other. Moreover, the complexity of the fields of action makes it impossible to foresee the consequences for the environment of any individual course of action.

It is the task of transdisciplinary research to connect fields of action with one another and to demonstrate the relationships between them. With every specialization goes its own language. To be able to communicate between the different fields and to make the problems evident requires the ability to speak in a language that lay persons can understand. Transdisciplinary research is a presupposition to be able to weigh the chances and risks of new technologies and of different terms of action against each other, to take over responsibility and feel at home in this world.

Dr. Ruth Baumann-Hölzle, Interdisciplinary Institute for Ethics in Health Care, Zurich

5

The Swiss Transdisciplinarity Award

5.1 Introduction: Goals and Criteria of the Award

by Heidi Diggelmann, President of the Swiss National Science Foundation, Berne; Gertrude Hirsch Hadorn, ETH, Zurich; Ruth Kaufmann-Hayoz, IKAÖ, University of Berne; Johannes R. Randegger, Novartis Services Ltd, Basel; Christian Smoliner, Federal Ministry of Science and Transport, Vienna, Austria

Abstract

Members of the Jury of the Swiss Transdisciplinarity Award explain the award's purpose, the selection procedure, and final decisions, supplemented by information about the Gebert Rüf Foundation, which financially supports the Award.

The Swiss Transdisciplinarity Award is an important incentive for transdisciplinary research. Conferred for the first time at the International Transdisciplinarity Conference, it was supported by a generous gift of 50,000 Swiss francs from the Gebert Rüf Foundation.

Three Challenges of Transdisciplinarity

Transdisciplinary research must tackle several quite different issues, which together are crucial for success:

- produce knowledge that is scientifically reliable
- address important issues in society to produce knowledge that is relevant for the future development of society and nature
- involve actors and stakeholders in society in stimulating the mutual learning process between science and society that is necessary to producing knowledge that is effectively used.

These different components are of equal importance and present equal challenges. The three components can be symbolized as the different parts of a pigeon (Fig. 1).

Figure 1
The Challenges of Transdisciplinarity Research are Symbolized as Three Different Parts of a Pigeon (Pablo Picasso, pigeon with varied flowers; © ProLitteris, 2000, 8033 Zürich)

The head represents scientific competence to produce reliable knowledge. The tail symbolizes issues in society that require relevant knowledge. The wings stand for the process of mutual learning by which knowledge is effectively used. All three components are necessary in order for the bird to fly.

Selection Procedure

The jury for the Award, comprised of the authors of this Introduction, are individuals from academia, science policy, and society. The jury decided to split the award into three parts, allotting 15,000 Swiss francs for each of three major components of transdisciplinary research (indicated below). The most risky and innovative of the final three winning projects was to receive 5,000 Swiss francs as a bonus. In addition, awards were to be conferred on the entire project team, not just the individual presenting the contribution at the conference.

Figure 2
Heidi Diggelmann, President of the Swiss National Science Foundation, with the Award Winners Egon Becker, Bernhard Truffer, and Robert Lukesch (from left to right) (photograph: keystone)

All submissions accepted for Mutual Learning Sessions, Dialogue Sessions and the Idea Market were eligible, with the exception of submissions by members of the Conference Board. Approximately 250 submissions sent for publication in the two pre-conference Workbooks were screened by members of the Conference Board, following defined criteria. For candidates to be considered, a written comment was required. The award committee classified the resulting pool of roughly 50 projects into the three categories of components. Subsequently, each contribution was assessed a second time in an independent evaluation by two members of the award jury. In a final meeting, the jury selected the top ten, then narrowed the field to three final winners.

Summaries of Winning Projects

The top three contributions received the Swiss Transdisciplinarity Award (Fig. 2).

- The award for achievements in the first category – theoretical issues – went to "Sustainability: A Cross-Disciplinary Concept for Social-Ecological Transformations," by Egon Becker, Thomas Jahn, Diana Hummel, Immanuel Stiess and Peter Wehling (*Chapter 5.2.1*). In a clear, profound, and paradigmatic way, this contribution addresses conceptual and methodological issues involved in the integration of analytical, normative and strategic knowledge.
- The award for the second category – issues important for the future of society and nature – went to "Ökostrom: The Social Construction of Green Electricity Standards," by Bernhard Truffer, Christine Bratrich, Jochen Markard and Bernhard Wehrli (*Chapter 5.2.2*). This project is an instructive example of how to address a political hot spot with scientific professionalism and political responsibility.
- The award for the third category – addressing participation of actors in society for mutual learning – went to "Research in Public – Suburbanization in Rural Areas," by Robert Lukesch, Wolfgang Punz, Helmut Hiess, Gerhard Kollmann, Franz Kern, Sepp Wallenberger, Bernhard Morawetz and Helmut Waldert (*Chapter 5.2.3*). An innovative and courageous approach to using media to involve stakeholders in transdisciplinary research, it also received a bonus of 5000 Swiss francs for risk.

Seven remaining contributions were in the final top ten

- "Vertisols and Ecohealth," presented by Mohammad A. Jabbar, M. A. Mohamed Saleem and Hugo Li-Pun (*Chapter 5.3.1*), effectively integrates scientific and indigenous knowledge as well as local conditions in resource management.
- "Search for Ecojumps," presented by J. Leo A. Jansen, Geert van Grootveld, Egbert van Spiegel, Philip J. Vergragt, Wilma Aarts and Conny Bakker (*Chapter 5.3.2*) offers a convincing concept for achieving sustainable technology research and development by backcasting from needs to products and from the future to the present.
- "Complex Systems Research," presented by Paul Jeffrey, Peter Allen, Roger Seaton, and Aileen Thomson (*Chapter 5.3.3*) is a paradigm for applying systems analysis to structure the complexity of cross-disciplinary knowledge in product design in the aerospace industry.
- "A Modest Success Story," presented by Pirkko Kasanen (*Chapter 5.3.4*) clearly delineates crucial elements for successful transdisciplinary research, especially participatory definition of goals and questions.

- "Innovative Urban Planning and Management," presented by Paulius Kulikauskas, Niels Andersen, Freddy Avnby, Lykke Leonardsen, Ole Damsgaard and Andreas Schubert (*Chapter 5.3.5*), is a lucid analysis of stakeholder participation and ways of overcoming obstacles in cross-sectoral urban planning.
- "Political Role of Participatory Technology Assessment," by Michael Nentwich and Danielle Bütschi (*Chapter 5.3.6*), provides illuminating answers to a crucial question. To what ends do we enter into participatory forms of research with the public, and what factors are important for success?
- "Urban Environmental Management," presented by Kaspar Wyss, N. Yémadji, A. Nodjadjim, Mamadou N'Diaye and Enda Graf (*Chapter 5.3.7*) demonstrates how to strengthen community initiatives in a participatory action-research approach, illustrated by a North-South research partnership for managing urban environmental issues in Africa.

Further information on the Swiss Transdisciplinarity Award is available at <transdisciplinarity@snf.ch> or at <http://www.transdisciplinarity.ch>.

• GEBERT RÜF STIFTUNG •

The Gebert Rüf Foundation
Philipp Egger, General Manager

Although the Gebert Rüf Foundation was founded as a private organization only two years ago, it has been involved in developing innovative ideas from the very start. This commitment is evident in its rapid acceptance by Swiss institutions of higher education. With an annual budget of 10 million Swiss francs, the Foundation can accomplish a great deal. It welcomes even large projects, which may otherwise have difficulty obtaining adequate financing. Its organizational structures are flexible, allowing it to make decisions about inquiries and applications efficiently and clearly. The structure of the management board and its members are a mark of the Foundation's professionalism. It is not only able to finance high-profile projects but is also closely involved with them every step of the way to ensure their effectiveness and success.

The Foundation regards transdisciplinarity as the most important criterion for developing innovative ideas. It is central to new teaching concepts and research methods, especially at the frontiers of formal science and specialized scientific institutions, and among everyone who uses a common learning process. Yet, "interdisciplinarity" and "transdisciplinarity," have become buzzwords. The Foundation conceived the idea of and endowed the Swiss Transdisciplinarity Award in order to confer greater prestige on the concept of transdisciplinarity and to promote substantive work in this area.

The award is intended to counteract powerful disincentives that discourage many exceptional people from daring to tread on the slippery surfaces of transdisciplinarity. The younger generation is especially at risk. Finding themselves labeled as "non-scientific," a label that damages their reputations, is a significant part of the problem. Academicians of all generations must relinquish the stance of splendid isolation within their own scientific worlds. They must also become involved with group processes that develop dynamically in ways that cannot be foreseen.

The Award also acknowledges the work of the Swiss Priority Program Environment (SPPE) of the Swiss National Science Foundation. Since 1992, the SPPE has fostered pioneering work in interdisciplinary organizations and research programs, especially in the debate about environmental resources and sustainability. The Transdisciplinarity Award represents a watershed in this history. For the first time in Switzerland there is an incentive to develop transdisciplinarity and for it to be understood by the international scientific community. This award can stimulate wider propagation of the idea of transdisciplinarity and its synergistic capacity. Hopefully, it can be made on a regular basis and will be a major contributing factor to influencing environmental sciences.

More information about the Gebert Rüf Foundation is available at <http://www.grstiftung.ch>.

5.2 Final Awards

The contributions of the top three recipients of the award are introduced by the formal statements of recognition the jury made when conferring the actual awards.

5.2.1 Sustainability: A Cross-disciplinary Concept for Social-Ecological Transformations

by Egon Becker, Thomas Jahn, Diana Hummel, Immanuel Stiess and Peter Wehling, Institute for Social-Ecological Research (ISOE), Frankfurt am Main, Germany

Recognition
The project "Sustainability" was chosen for its analytically sound and comprehensive design of transdisciplinary research on sustainability, which resulted from a participatory process involving scholars from North and South.

The Problem

Since the 1992 Earth Summit in Rio, sustainability has become an internationally accepted keyword for a political discourse committed to quality of life, conservation of natural resources and a sense of obligation to future generations. Rather than being a well-defined concept, sustainability might best be characterized as a contested discursive field that allows for articulating political and economic differences between North and South, as well as a concern for social justice and political participation within environmental issues. Moreover, in a broader sense, the worldwide discourse on sustainability can be envisaged as a rallying point of public debate, knowledge-building practices and political strategies for coping with a series of unprecedented global and regional problems.

With growing public awareness, sustainability discourse has also become a "generator" of a new and innovative type of research issue and is closely linked with economic and political problems. At the cutting edge of sustainability-related research and scholarship, innovative approaches have emerged. They constitute a new transdisciplinary field, with characteristic problems. Yet, both social and environmental-oriented natural sciences are poorly equipped to cope with this new set of problems that cut across existing disciplinary boundaries. Theoretical concepts and analytical tools are developed mainly within given disciplinary matrices. Despite considerable attempts by both the social science and the natural science communities, an analytical framework for cross-disciplinary research oriented to social actors and social-ecological transformations is still missing. Social science approaches, especially, have not been incorporated seriously into the mainstream of environmental research. And vice versa: to a large degree environmental considerations are excluded from the mainstream of social science. Moreover, there is a deep gap cutting across social sciences themselves.

The Project

In response to this situation, an international project on "Sustainability as a Concept for the Social Sciences," was launched by UNESCO within its intergovernmental research and policy program, MOST (Management of Social Transformations). From the very beginning, the project was designed and organized by the Frankfurt Institute for Social-Ecological research (ISOE) on a very low budget.

At an *organizational* level, the task was to build a viable international network of scholars from the North and the South who represented the full range of social science disciplines. In the end, a gender-mixed group of about 15 competent scholars from different branches of social sciences, and with various regional and cultural backgrounds, was established. The study group assessed the current state of the debate on sustainability and its future prospects in their respective fields.

At a *cognitive* level, the aim was to explore conceptual and methodological implications of sustainability for the social sciences and to outline a framework for cross-disciplinary sustainability research. Additionally, in the course of this project, participants came together at a conference held at Frankfurt am Main in November 1996. The conference produced numerous papers (Becker and Jahn 1999), and a first synthesis of the results was published as a MOST-Policy-Paper (Becker et al. 1997).

The Outcome

The project was a kind of academic adventure. Looking back, its success was the result of two procedural assumptions. At the organizational level, we worked strictly with a participatory round-table model, together with strong conceptual inputs and common evaluation of outcomes. At the cognitive level we tried to avoid the "fallacy of misplaced concreteness," to quote the philosopher Alfred North Whitehead. The study group devoted a lot of energy to generalizing and clarifying the different conceptual approaches introduced by members of the group.

Integration of Actors, Technologies and Knowledge

To overcome the weakness of discipline-oriented research at the core of sustainability work, the study group identified and tackled a twofold challenge of integration:

- The first challenge results from the hybrid character of the sustainability concept itself. It is simultaneously a political model for change with a strong normative content and a concept for scientific analysis. Therefore, in sustainability research different academic and non-academic actors with heterogeneous interests and knowledge interact with one other, creating a need for coordination and *social integration* within a common frame of action.
- The second challenge results from the differentiation of scientific and technological knowledge across a broad spectrum of disciplines and practical applications. Within the field of sustainability research, the well-known problem of interdisciplinarity is increased by the deep gap between social and natural sciences. Therefore we need a kind of *cognitive integration* to bridge this gap.

Distinctions within the Discourse on Sustainability

Integration demands differentiation. Hence, we distinguished three levels within the discourse on sustainability: normative, operational and analytical.

First, the idea of sustainability introduces a new set of *normative* commitments into the long discussed problematic of development or modernization, as well as the environmental debate. A call for justice is being made on behalf of future generations. Societal development should on no account lead to irreversible constraints on the chances of future generations meeting their needs. In addition, international justice between North and South, social justice within societies, equity in gender relations and democratic participation in decision-making processes are widely discussed normative aspects of sustainability. How, though, can we attempt a kind of international and intercultural agreement on those ambitious normative commitments?

The most grating controversies occurred on the normative level. Instead of setting definite goals for action in defined situations, we tried to work out, by means of a norm-analysis, "meta-norms." These general meta-norms were capable of directing the process of norm-building with different actors and in different situations. Examples include "capability of processes to continue," "enhancing the potential for development," and "co-evolution of social and natural systems."

Second, at the same time sustainability imposes a strong commitment to action aimed at change directed towards reshaping relationships between human beings and their environment. This occurs along different dimensions, reaching from patterns of production and consumption to reproductive behavior relevant for demographic transitions. At an *operative level* different concepts of action for the fields of politics, economy or culture must be developed, on the basis of general criteria of sustainability. Another question, then, arises. How can we work out, together with key actors in different fields of society, practical concepts for goal-oriented actions?

An important step is to strengthen the actor-orientation of research in order to assist in developing strategies that enhance the ability of key actors to move towards more sustainable practices. This step implies integrating the knowledge held by non-scientific actors within the research process. Actor-orientation of research could entail, in particular, choosing topics for investigation according to the needs of key actors, inviting non-scientific users to reformulate research issues, and involving them in various stages of the research process.

Third, at an *analytical level*, sustainability claims that societal development can no longer be viewed without considering its natural prerequisites. It is inseparably coupled with their reproduction. The focus of research problems, therefore, will be on transformations of patterns of interactions between societies and their natural environment. How, though, can we analyze these transformations?

The project tackled the most complicated problems at the analytical level. Achieving a more comprehensive understanding of interactions between societies and their natural environment generates the hard core of the cognitive problem of integration, mentioned previously. It is located mainly within scientific discourse. The new knowledge base for sustainability requires innovative tools and strategies for integrating knowledge created in different branches of social science. Even more challenging, sustainability research needs bridge concepts between social and natural sciences. Sustainability holds far-reaching implications for the cognitive structure of social sciences and environmental research, the latter of which is still dominated by natural science approaches.

Social-Ecological Transformations

The international discourse on sustainability is centered around a new type of recently evolved complex global and regional problems. They result, on the one hand, from economic and technical globalization and their social, political and cultural impact. On the other hand, global and regional environmental changes are often intertwined in an indistinguishable manner with the globalization process. In the study group, we experienced how societies in the North and the South perceive these problems in different ways, in light of growing public awareness and refined scientific descriptions. They define problems within their prevailing institutional and intellectual frameworks, tackle them with available methods, solve them partially, and create new and previously unknown subsequent problems of a second order.

To sum up: In sustainability research science is part of the problem and part of the solution. Scientific efforts are embedded in a dynamic, self-referential process of solving and creating social and ecological problems on different scales of space and time. In these processes patterns of societal relationships to nature undergo major changes. Thus, we call these changes *social-ecological transformations*.

These transformations can be produced either by non-intentional evolution (such as economic globalization or spread of new technologies) or by goal-oriented innovations and coordinated management activities. Mostly, though, they are out of social control. With the normative concept of sustainability, desired changes are marked. Yet, how we stimulate or manage them at the operative level remains an open question. Surely we need knowledge about the behavior of strongly coupled social and ecological systems containing feedback loops – a type of *complex systems* we are only beginning to understand.

At the analytical level the term sustainability was used within the UNESCO-project mostly with a negative sign of absence, in order to identify non-sustainable states and processes. Analytically, this opens a "corridor" for different paths to more sustainability, a corridor bordered by "crash barriers." Sustainability is related to stabilized and preserved patterns within social-ecological transformations over long periods of time. We addressed sustainability problems strictly in terms of processes instead of state. Following this procedure allowed us to redefine existing definitions of sustainability.

Conclusions

Fostering integration of knowledge requires new forms of cognitive and organizational alliances which need the support of research policy. The initiative of MOST – to bridge the gap between sustainability as a political framework for action and as a framework for scientific research – was a crucial prerequisite for success in the project. Additionally, we learned that transdisciplinary research requires an appropriate framework both at cognitive and organizational levels. Besides strict methodologies and sophisticated conceptual schemas, transdisciplinary research requires new forms of cooperation and networking, such as inclusion of researchers and laypersons with different social, cultural and gender experiences. The better these two levels are coordinated, the faster the onset of positive effects on theoretical and practical problems.

5.2.2 Ökostrom: The Social Construction of Green Electricity Standards in Switzerland

by Bernhard Truffer, Christine Bratrich, Jochen Markard and Bernhard Wehrli, Swiss Federal Institute for Environmental Science and Technology, Kastanienbaum, Switzerland

Recognition

The project Ökostrom "Green Electricity Standards" was chosen for addressing a political hot spot in Switzerland with professional interdisciplinary research as well as a regard for political and economic realities and necessities.

Introduction

Markets for green products have received considerable attention in many countries recently. Organic food, for instance, is gaining increasingly important market shares, and other product sectors are following suit (Villiger et. al. 2000). The liberalization of electricity markets has also opened up opportunities for electric power companies to differentiate their products in the marketplace. Conventional electric utilities, as well as new Green Power marketers and traders, have begun to develop Green Power products (Loskow 1998; Holt 1997). As in other Green Goods markets, quality control is a key success factor. Recently, a number of Green Power certification initiatives have emerged, aimed at assessing and communicating environmental benefits of Green Power products (Markard and Truffer 1999). Worldwide, about a dozen labeling schemes exist today.

Hydropower has an ambivalent position among electricity generation technologies, which are included in Green Power products. This ambivalence is reflected in the variety of criteria used to identify Green Hydropower plants. Most labeling schemes set a limit at generating capacity. Others restrict admission to old plants and do not want to support new projects. However, from an environmental standpoint, neither capacity nor construction date has ecological significance. An environmentally-motivated and scientifically-derived set of criteria would correspond better to the needs of a trustworthy eco-label.

Political Debate on Green Hydropower in Alpine Countries

One major reason for the ambivalent treatment of hydropower is associated with environmental characteristics of hydropower plant operation. Electricity from hydropower plants creates no air pollution, no nuclear waste and, in most cases, virtually no CO_2 is emitted. Furthermore, hydropower plants are extremely energy-efficient, and energy from storage plants is available instantaneously. Nevertheless, construction and operation of hydropower plants have local impacts, which may be severe. They include extinction of fish populations, loss of aquatic habitats, sinking ground water levels, and deterioration of landscapes.

This double characteristic – a globally preferable and locally-destructive energy system – has gained high political relevance in many countries (IUCN and World Bank 1997). The negative effects are pervasive, especially in the alpine region of central Europe. Most rivers have been dammed for hydropower generation. In Switzerland, this situation led to considerable political struggle (Truffer 1999). In the 1980's conflicts escalated in the form of public protest movements against new dam projects. As a result, a new and stricter water protection act was passed by popular vote in 1991, deepening political segregation into two conflicting camps – hydropower producers and environmentalists.

Although the water act demands environmentally-optimized operation modes, few plants have adopted them. Long-term licensing of water use rights for power plants is a major reason. The renewed water protection law demands, for instance, minimum flow requirements. But, its application is tied to relicensing of power plants. For the majority of plants, this renewal is due in 20 to 40 years from now. Furthermore, market liberalization has changed the economic context of hydropower generation dramatically. Most Swiss dam projects have been abandoned or postponed, and some existing hydropower plants are threatened by closure if no supporting actions are taken.

Defining New Roles for a Research Institution

The Green Hydropower Project

Analysis of the political background and future challenges of the hydropower sector led a team of social and natural scientists to start a transdisciplinary project aimed at developing an eco-label for Green Hydropower plants (Truffer et. al., 1998). Members of the team are primarily located at and supported by the Swiss Federal Institute of Environmental Science and Technology (EAWAG). The project was started with the explicit task of developing a "new kind" of research.

For three decades, EAWAG has been successful in advising governments about technical water protection measures, such as waste water treatment plants and lake oxygenation. Additionally, it has established an international reputation for high-level basic environmental research. Research and practical applications have developed alongside each other for a long time. Increasingly, however, we have realized that by continuing to develop ever more sophisticated end-of-pipe technologies and by mainly increasing output of research results, the institution is losing its grip on newly-emerging problems.

These new problems are defined by complex cause-effect relationships that demand increased collaboration between different disciplines in environmental sciences and engineering. Furthermore, solutions must be developed for complex institutional settings. Therefore, social sciences have been identified as potentially important research partners. Finally, solutions must be defined by explicitly taking value considerations into account. This means projects have to be organized to allow stakeholders stronger involvement than in the past.

Given the lack of scientific background in hydropower certification and the highly politicized context of hydropower regulation, the project team identified development of an eco-label for Green Hydropower as an interesting focus. The team began by carrying out extensive interviews and discussions with hydropower operators, government officials, and environmentalists. As a result, the project could be substantially focused and a first constituency formed.

Both plant operators and environmental organizations emphasized the urgent need to find new and less antagonistic ways for dealing with the public image and environmental impacts of hydropower plants. Green Hydropower was seen as a potentially beneficial but highly uncertain opportunity. Additionally, it was clear that none of the traditional conflict groups could take the lead in trying to overcome the dilemma. By proposing a scientifically-grounded Green Power certification procedure, the research institution could play an important role in supporting a more rational discussion about options for sustainable hydropower production.

These findings led to defining the following goals:

Project Goals

1. A certification procedure for hydropower plants had to be developed that would respect the high complexity of ecological and political contexts of hydropower operation. This procedure should be adapted to the specific political and ecological context of Switzerland, but also be applicable in an international (at least European) setting.
2. The substantial uncertainties with regard to future market conditions of the green power products should be reduced, as far as possible, by defining a number of social science research projects focusing on these topics.
3. A broadly-based constituency should be put into place, capable of making decisions about highly value-laden questions and supporting the eco-label scheme.
4. Finally, the whole project should create a context in which high-level social and natural scientific research projects could take place.

At the beginning of the year 2000, the Green Hydropower effort encompassed about 20 individual projects, divided into four working groups that tackle specific problem areas:

- One group focuses on developing new tools for determining minimum flow regimes. It encompasses projects on fish and benthic ecology and its interaction with habitat and hydrology, sediment transport, temperature regimes and chemical substances.
- A second group analyzes adverse effects of hydropower operation on flood plain ecosystems. Exchange processes between ground water and riparian ecotones, distribution of benthic organisms and terrestrial vegetation, and the importance of sediment transport are analyzed.
- Results of these two groups are fed into the third group, focused on assessment. This group, which develops a tailor-made certification procedure, must specify a scientifically sound, credible and effective procedure.
- The fourth group concentrates on market and political sides of Green Power products. It develops marketing strategies, analyzes learning processes of consumers and looks at complementary policy measures to enhance the effectiveness of Green power markets.

A pilot study of the project was carried out during the first year, with the goal of setting up the project and finding support from major stakeholders. Work began with a two-year case study, focused on options of sustainable hydropower operation in the Blenio valley in the southern Swiss Alps. The goal of this phase was to develop a comprehensive assessment procedure, applied to the first hydropower plants in summer 2000. Following this phase, scientific projects will focus on specific research issues and on certification of concrete hydropower plants.

Lessons Learned

The product of the Green Hydropower project – the eco-label – is only one result. At a more fundamental level, development of widely-accepted standards is at stake. Two questions are uppermost. How is a system of environmentally sound electricity generation defined? And, how could we describe and reach a preferable energy future? Such standards cannot be prescribed by experts alone and do not derive from purely objective facts. They will be socially constructed in processes of mutual learning by and among diverse societal actors.

A research institution striving for objectivity and truth still has an important role to play. In our case, it must provide tools and methods for assessment of river ecosystems used for hydropower generation. Researchers have to structure different aspects of hydropower management and to communicate relevant facts that can support rational decisions. However, they must be prepared to define their role in an ongoing learning process and to clearly communicate the limits and uncertainties of their knowledge. They must also be explicit about where value judgments should be made by different interest groups. The Green Hydropower project has yielded some initial success in actualizing new roles, making it a promising though highly demanding undertaking.

The promising part is that all involved parties contributed essential elements to solving the problem: environmental scientists defined a scientifically sound assessment procedure and developed tools to evaluate and simulate effects of environmentally optimized operation modes. Social scientists aimed at reducing uncertainties related to market liberalization and at sketching strategies for Green Power market development. Representatives from different stakeholder groups had to overcome taken-for-granted problem-and-conflict definitions and to embark on a mutual-learning process.

The demanding part was overcoming a number of prejudices among the researchers, in two respects. First, the need for a professional project management and a strong product orientation created a lot of discussion and irritation among many scientists. One major conflict was the differing time horizons of scientists and stakeholders. Scientists regarded time horizons of at least five years as a starting point for involvement. Stakeholders, on the other had, did not show interest in a common undertaking that would take more than a couple of months. Ironically, it took almost three years for a common standard to develop, a risky time lapse for both parties.

Second, many stakeholder groups did not accept the scientific institution as a "neutral" mediator. Instead, the project team was accused of being an agent of the "other" side of the political conflict. Building trust, therefore, required recurrent bilateral exchanges with several representatives of all interest groups. These time-consuming processes could only be carried out because of strong moral and financial support from the directorate of EAWAG. Several times the project was on the verge of being closed because of heated opposition to the initiative.

Taking these difficulties into consideration, it is clear that transdisciplinary research in a highly politicized problem area is not an easy road to travel. Problems are extremely complex, uncertainties are substantial, and the probability of failure is high. This is all the more true when the ultimate success or failure of a project depends on the good-will and initiative of stakeholders and is, therefore, outside the control of project management. One of the most crucial variables is the ability to translate between different positions, both within stakeholder groups and among scientists from various disciplines.

At the same time, transdisciplinary research has shown some definite benefits. First, researchers, especially young students, become highly motivated to work in a context that shows immediate relevance. At the level of individual researchers, a deeper understanding of the variety of aspects of a particular problem can be gained. New solutions may also be developed, a prospect that rarely emerges when following purely disciplinary research routines. Additionally, exchange with different stakeholders proved to be inspiring and motivating for researchers.

The ultimate contribution of an eco-label for electricity to resolving environmental problems is still too early to assess in detail. The instrument holds promise, but in the end the market has to tell. Independent of ultimate market shares that may be reached by eco-friendly power products, the eco-label definition process has created a context in which older conflict positions can be questioned and space for compromises built. All in all, we experienced the project as a highly motivating and gratifying experience and are eager to evaluate it in a way that increases potential benefits while decreasing difficulties in future transdisciplinary undertakings.

5.2.3 The Green Leaves of Life's Golden Tree: The Project "Research in Public"

by Robert Lukesch, Wolfgang Punz, Helmut Hiess, Gerhard Kollmann, Franz Kern, Sepp Wallenberger, Bernhard Morawetz and Helmut Waldert, ÖAR Regional Development Consultants Ltd, Fehring, Austria

Recognition

The project "Research in Public" was chosen as an innovative and courageous participatory approach to developing a sustainable way of suburbanization in rural areas.

The Topic: Suburbanization in Rural Areas

- *Suburbanization* is a ubiquitous phenomenon that we address with regard to challenges for rural areas. The project "Research in Public" was carried out from the focus of cultural landscape research, which the Austrian Ministry of Science has promoted and financed since 1995 (BMWV 2000).
- *Cultural landscape* carries not only essential life functions and provides space for human activities. It also conveys an aesthetic context of quality of life and one of the most influential image factors in the minds of Austrians and foreign visitors.
- *Austrian rural areas* are an outcome of historical striving for survival among farming populations. Having absorbed a continuous flow of clearing, cutting, mowing, sowing and harvesting labor through generations, farmers today are sanctuaries of Austrian identity.

While generally acknowledged, these facts are contravened by a lax attitude evident in the increasing consumption of open landscape. Settlements are spreading and commercial activities being dispersed to peripheral zones called *edge cities* (Garreau 1991) or *para-cities* (Holzinger 2000). Both growth phenomena can been criticized:

- Typical landscape features are degraded by "urban blight" (Zibell and Gürtler-Berger 1997).

- Urban characteristics get lost in their peripheries (Pawley 2000).
- Resources such as soil and fuel are wasted and pollution increases, mostly due to higher energy consumption for traffic and dispersed housing (Schremmer 1999).
- The financial burden rises for constructing and maintaining extensive supply infrastructures (Schremmer 1999).
- Commercial activities in urban centers decrease, leading to loss of workplaces and thinning of vicinity services (retailers and service shops next to where people live)
- Society becomes fragmented and sometimes partially isolated in peri-urban ghettos (Heise-Verlag 2000).

Many parties are involved. *Space planners* and *local authorities* try to stop these trends through authoritative or contractual regulations (Örok 1995). However, regulations sometimes seem to cause more chaos than they prevent (Zibell and Gürtler-Berger 1997). *Municipalities* compete with neighboring communities for tax benefits from investments on the green lawn (investments outside settlement limits, or along exit roads) (WKSt 2000). Some attempt intermunicipal cooperation, but successful examples are rare. *Entrepreneurial lobbies* undertake city marketing initiatives to countervail centrifugal forces, fostered by the attractiveness of extensive space for productive or commercial activities and the generalized use of cars. *Consumers* appreciate extensive parking lots, supermarkets' competitive offers, and the convenience of buying everything under one roof. Yet, not everyone owns or has access to a car, especially the elderly. Young people are attracted by peripheral movie centers and hangouts, but they resent the decline of gathering places in their home towns and villages.

The shady sides of suburbanization take longer to surface than their immediate image-boosting benefits. Negative effects are often masked by dilution or dislocation:

- Exploding infrastructure costs such as waste water devices are subsidized by federal funds (Schremmer 1999) and not fully carried by those who cause them.
- Traffic congestion may disturb a neighboring community more than the immediate location of a commercial center.

As we see, suburbanization in rural areas is complex in its temporal, social and spatial dimensions:

- Its pace has accelerated in recent times, as economic internationalization moves forward and tears down barriers of geographical competition.
- Phenomena related to suburbanization processes touch nearly everybody in some way.
- It is an issue of "general interest."

The effects on cultural landscape are deep, hard to reverse and in direct conflict with long-term sustainable development. This multifaceted problematic requires not only an interdisciplinary but a genuinely transdisciplinary approach. Shifts of perception go hand in hand with social and political action, and scientific observation cannot be separated from introducing changes either. Complex problems require complex answers (Malik 1992).

Project Setting

We chose two project sites, the rural towns of Horn (Lower Austria) and Weiz (Styria) and their neighboring municipalities. Our *first hypothesis* was that Weiz (9245 inhabitants, 5 km^2) does better in dealing with the described problematic than Horn (6364 inhabitants, 39 km^2). The reason appeared to be intermunicipal cooperation within the territorial association REV (Regional Development Association), which has 17 member municipalities in the Upper Raab valley between the rural towns of Weiz and Gleisdorf.

This led us to a *second hypothesis*, that creation of cooperative structures depends on essential factors which either promote or hamper their emergence and consolidation. There is, for example, greater open space in Horn, which is nearly eight times larger than Weiz and has fewer inhabitants.

The project is also embedded in ongoing local development processes:

- An eco-management plan being implemented in Weiz (Ökologie-Institut 1996, 2000, Ökoplan Weiz) and first attempts to coordinate strategic resource management on the intermunicipal level within the REV.
- A *city marketing initiative* in Horn, whose urban center suffers severely from business dispersion to the city limits, known popularly as the "golden mile."

The team has the following members:

- the coordinator (a regional development consultant)
- a team of space planners
- a team of ecologists
- a team of social scientists
- a marketing specialist
- two local coordinators (integrated into local actors' networks, one the manager of a local innovation center in Weiz, the other a city marketing manager in Horn)
- an Austrian broadcast journalist, who links up with colleagues in local radio and TV.

Aims

The project has four aims:

1) *Shifting perception*: We want to raise awareness among local populations and interested consumers of broadcasting and TV programs of the effects of private and job-related roles in the suburbanization phenomenon. They will gain insight into the ambivalence of behavior patterns stimulated by short-term and economic benefits that are not generalizable.
2) *Creating strategic options*: We want decision makers and economic key players to perceive additional choices for developing the infrastructural and economic context, based on shared visions.
3) *Mutual learning*: We want key actors in both locations to have an opportunity to exchange experiences and to transfer good practices in problem solving.
4) *Making language permeable*: We want to gain new insights into possibilities of interlinking scientific and everyday language, partially mediated by mass media, for societal problem solving.

The fourth aim, in particular, is driven by the fact that we deal with three types of "*public*," and, therefore, three different languages: the languages of researchers and technicians, of local actors and decision makers, and of the "broad public" represented by mass media consumers and journalists.

Methods

The methods respond to two main questions:

- How can we make the creeping spatial processes visible and tangible for the population, usually preoccupied with other matters and not perceptive about those changes, in spite of their high esteem for an intact cultural landscape?
- How can we connect local problem solving with the more abstract concept of sustainable development?

In order to meet the *first requirement* we chose the method of *future images and stories* (BMWV 1998). *Hiess* defines the method:

> *This method has been elaborated and tested in a number of cultural landscape research projects in Austria. It intends to enhance the awareness towards potential changes, to strengthen the motivation for common action and to improve the transparency of consequences of supraregional decisions on a regional level. To this end, regional images of the future will be developed on the basis of supraregional sectoral scenarios. By combining utilization of maps, aerial photographs and fotos into cut- and paste montages, the landscape is depicted at a future point in time. At the same time, the everyday life of various protagonists (farmers, tourists, mayors, entrepreneurs, etc.) is to be described at a future moment based on foreseeable developments. In addition, the possible impact on the landscape will be deducted from their behavior.*

In order to meet the *second requirement*, we chose an interactive, team-oriented instrument for assessing the territorial potential for change towards sustainability. Its roots lie in the empirical evidence of several hundred case studies on innovation in European rural areas, analyzed by a working group of the European Rural Observatory (AEIDL 1997). The instrument of *innovation compass* was developed by the ÖAR Regionalberatung (Lukesch 1999) as a result of inductive conclusions based on that empirical work. It combines regional diagnosis with identification of key development strategies. The compass comprises nine realms of territorial development:

a. Environment (natural and built heritage and resources)
b. Images (people's own views of the territory, other people's views of territory, and people's assumptions about the views of others)
c. Markets (internal and external exchanges of goods and services)
d. Financial resources (financial capital and commercializable assets)
e. Enterprises (profitable and non-profit activities)
f. Competences (formal and tacit knowledge and skills)
g. Governance (rules, institutions and decision making processes)
h. Identities (feelings and beliefs about the collective "self")
i. Human Resources (demographic features, human potential and relations)

The results of rating these nine components in a team process are visualized in a nine-edged radar whose respective shape allows deducting hints for appropriate development strategies under the guiding principle of sustainability.

Description and Actual State of the Process

The research project has three overlapping phases (Fig. 1):

First Phase

The project started with a detailed diagnosis of current socio-economic and ecological states and trends on project sites. Relevant elements of general sectorial trend forecasts are put in relation to the local situation, in order to create a coherent future image of the place around the year 2025. These future images are synthesized in an *expert workshop* with the research team and local key actors and decision makers. This workshop marks the start of the *mobilization phase*, during which quantitative elements are complemented by individual "future stories" and other creative local inputs (from local people, schools, etc.). Receivers of directly mailed leaflets and readers of these posters are invited to express an opinion on the possibly irritating or worrying pictures. They are also invited to participate in Future Workshop 2025, two months later, where an alternative, desirable future is elaborated. Local and Austrian-wide broadcasting contributes to spreading this call for participation and contributions.

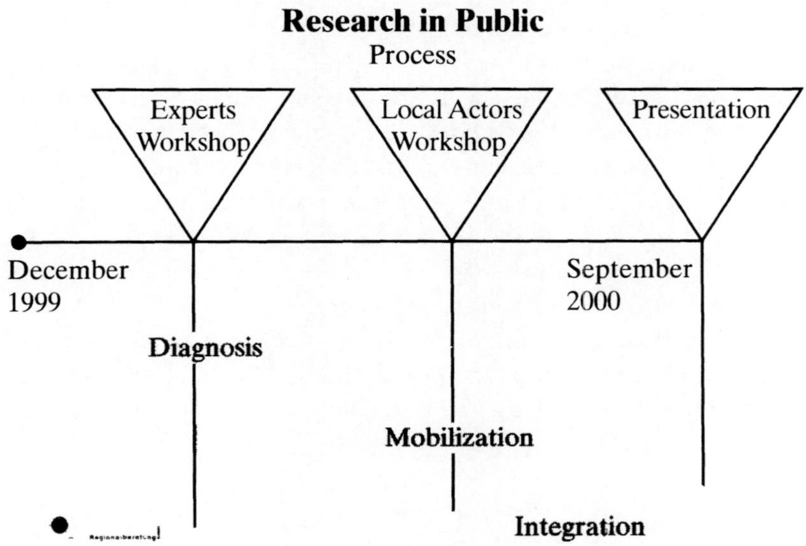

Figure 1
Process of Research in Public

Second Phase

After duly screening and selecting suggestions, the material (future images, stories and other contributions) are exhibited. The above mentioned Future Workshop 2025 results in a shared vision of a desirable future with regard to specific themes (such as traffic, urban revitalization, intermunicipal cooperation, etc.). This Workshop is the most important event of the second phase, marking the end of the diagnosis phase and the start of the *integration phase*.

Third Phase

The thematic issues are concretized within work groups until final results are presented in each area, two months after the Future Workshop 2025, with key actors of other project sites present. The final presentation marks the end of the mobilization phase and the project. We presume the third phase (*integration*) will go on beyond the project, and its results will be embeddded into local development strategies. Results will be broadcast in an Austrian-wide program.

The project started in December 1999. Expert workshops and future workshops have been held with encouraging outcomes. Municipalities and local businesses are willing to support implementation with their resources. Scholars' workshops are held because they are eager to learn how to make a radio transmission together with the journalist. The topic will be – of course – the future of the place in which they live. Furthermore, people contributing suggestions receive vouchers for shopping in the urban center, contributed by local retailers.

The final presentations will be in fall of 2000. After this event, the expert team will step out of the process, but local people will continue to build the world in which they want to live.

5.3 Finalists

The contributions of the remaining seven finalists are introduced by the formal statements of recognition the jury made when conferring the actual awards.

5.3.1 Evolution Toward Transdisciplinarity in Technology and Resource Management Research: The Case of a Project in Ethiopia

by Mohammad A. Jabbar, M. A. Mohamed Saleem and Hugo Li-Pun, International Livestock Research Institute (ILRI), Addis Ababa, Ethiopia

Recognition
The project "Integrated Resource Management: Vertisols and Ecohealth" was chosen for effectively integrating scientific and indigenous knowledge as well as local conditions in resource management.

Introduction

Development is a human problem aimed at changing humans and their economic, social, and ecological environments. Throughout history, humans have acquired new knowledge to shape their future. As their needs multiplied and became more complex, science and technology – processes of systematic inquiry for acquiring and applying knowledge – have exerted increasing influence on human life, society, and the environment. To address the complexity and multiplicity of problems, scientific research has been divided into disciplines by problem areas. It has also been divided into basic, strategic, and applied or adaptive orientations, depending on whether the objective is to search for new horizons in knowledge or to adapt a known technology in a different situation. These changes have resulted in divisions of labor in knowledge production and application, and they have contributed to a phenomenal increase in output, income, and human welfare. This chapter focuses on a particular project, examining the limitations of disciplinary research to solve complex problems and the potential role of transdisciplinary research to overcome them.

Need for Transdisciplinary Research

The achievements of disciplinary research have been accompanied by negative outcomes, such as environmental degradation, economic disparity and continuing food insecurity. These outcomes have led to the realization that "agricultural activities take place within a complex mess of multi-scalar, multi-dimensional interactions"; the solution to any problem, as a result, should be sought through multidisciplinary or interdisciplinary systems approach in education and research (Lipton 1970; Epstein and Scarlet 1975).

Unfortunately, the talk/action ratio in systems research has been high for several reasons:

(a) lack of professional prestige in systems research
(b) the time-consuming and expert-intensive nature of systems research
(c) the reluctance of scientists to accept the need for crossing disciplinary boundaries to address complex problems.

Lack of adequate progress in systems research has raised doubts about the socio-ecological sustainability of science-based agriculture and agricultural communities in an age of increasing globalization. It is also increasingly recognized that improving human well-being and promoting sustainable and convivial human communities requires a holistic and integrated approach that transcends disciplines. Moreover, the people making key decisions about technology and policy choices and their outcomes must participate fully in the research process alongside formal researchers, not merely as recipients of research results. Transdisciplinary research starts from the premise that any problem or complex reality can be viewed and interpreted from a variety of non-equivalent perspectives; within each perspective a problem or reality can be understood from a range of spatial and temporal scales (Rosenfield 1992; Smit et al. 1998). For example, the phenomenon of global warming can be described and understood from different disciplinary perspectives at global, regional, and local levels over varying periods of time. Within each perspective and scale, stakeholders contributing to global warming and those suffering the consequences may identify different elements of importance, use different indicators, and draw different conclusions. Whose perspectives are considered, how they are incorporated in research and the scale of research all determine the outcomes. Transdisciplinary research may help in integrating various perspectives and scales.

Problems of Highland Ethiopia and Some Related Research

Poverty, malnutrition, low crop and livestock productivity, and resource degradation are major problems in rural East African highlands. They are highly interrelated. They reinforce one another, keeping the rural population in a vicious cycle of underdevelopment and environmental degradation. In the Ethiopian highlands, population pressure has pushed cultivation and livestock grazing to steep slopes and fragile lands, causing serious devegetation and soil erosion. Yet, about 12 million ha of Vertisols (heavy though fertile soil) remain underutilized because of poor internal drainage and resultant flooding and waterlogging during the rainy season. To avoid waterlogging, some farmers plough fields before and sow seeds towards the end of the rainy season. That way crops grow on residual moisture, producing low yields and exposing bare soil to heavy rains that cause erosion.

In food-deficit Ethiopia, removing constraints to crop production in Vertisol areas is an important entry point. A consortium of national and international research centers developed a technology package composed of an animal-drawn equipment called "broadbed maker." This package facilitates draining water during heavy rains, higher-yield wheat varieties for early planting that extends the growing season, and appropriate input and agronomic practices.

Indigenous knowledge and farmer preferences were considered in designing the package, which was tested in selected farm-sites with farmer participation. Later, development agencies diffused it in wider areas. A yield of 2–3 t/ha (tons per hectare) of wheat, compared to under 1 t/ha with traditional technology, plus an early harvest was expected to relieve the food security problems of poor farmers. Economic analysis showed significantly higher profit and employment. Adoption analysis indicated that farmer knowledge, capacity, and incentives were important at every step of technology generation and diffusion.

Technology adoption is not one-directional. Farmers move from acquiring knowledge to adopting it in continuous or discontinuous use.

In some cases, technology created externality, when water drained from plots on upper slopes created waterlogging in plots downstream. Solving such problems required community involvement in watershed management. This hypothesis was tested in a pilot watershed where common main and subsidiary drains were constructed with the voluntary participation of farmers. The process involved water management, drainage technology, and farmer-organization related research. Individual households' contributions in a collective effort to create common goods were positively influenced by the potential benefit to participants.

Inadequate and poor-quality feeds are major reasons for low livestock productivity. Given the scarcity of land, strategies were sought to increase both food and feed production in a complementary way, by integrating food and forage crops, multipurpose trees, better utilization of feeds and cycling. Research on feed production included selection of potential forages based on their environmental adaptation, feed quality and resource needs, and integration of forages in various cropping patterns (inter-, relay, and alley crops). In some cases, cereal-grain yields increased in association with forages, though greater benefit occurred in the amount of feed per ha. These studies also showed higher water-use efficiency and better nutrient cycling.

Economic analysis showed that, compared to pure cereal stands, crop-forage intercropping significantly increased gross margin and cash income. Returns were further enhanced when combined with crossbred cows for milk production. Several multipurpose trees suitable for different altitudes showing different attributes (such as frost tolerance and growth rate) were identified through participatory on-farm testing. Through farmer-to-farmer diffusion and seed sharing, many more farmers in and around the original research sites have planted these trees.

Seasonality and inter-year variability in feed quality and quantity aggravate feed problems. To better utilize available feeds, on-station and on-farm studies were conducted with crossbred cows for milk production and traction. The need for oxen and additional feed was reduced. Results also showed that with adequate feed supplementation, crossbred cows could be used for dual purposes, since the minimal power requirement of small farms did not significantly affect milk yield and reproduction. Moreover, crossbred cows significantly increased cash income and household nutritional status, especially of pregnant women and children.

Soil fertility is declining because manure is principally used as fuel and chemical fertilizers are expensive. Efforts to improve soil fertility include livestock production, efficient use of crop residues and manure, and introduction of herbaceous/tree forage legumes that can fix atmospheric nitrogen. Trials on the effects of grazing pressure showed that where manure was left on grazed plots, with grazing pressure, biomass productions increased and soil erosion diminished. Where manure was removed from grazed plots, biomass production declined and soil erosion was well in excess of the permissible limit. Moreover, feed shortage could be averted by synchronization of grazing with seasonal herbage availability at different slopes. Strategic fertilizer application could also improve biomass productivity and protect soil.

From Vertisols and Watershed Management to Integrated Resource Management

The component technologies described above were tested initially at the levels of plot, animal, and farm. Their economic viability was tested separately and partially in terms of yield and income. It was, though, a multidisciplinary project. Some farmers participated in several interventions, but their integrated effects have not been properly assessed. Improvement of human welfare – poverty alleviation, food security, better health and nutrition – and conservation of natural resources are ultimate research goals for the consortium. Assessment of the impact of various interventions should go beyond estimating profitability and productivity to explicitly considering final goals as assessment criteria. Inter-relationships between biophysical and human dimensions also need to be integrated both spatially and temporally, in order to identify ways to improve conditions of ecosystems and human welfare. This effort requires human, policy and technical dimensions to be integrated at household and watershed or community levels. To achieve this goal, the project is currently using agroecosystem health as a framework for analysis and synthesis that incorporates an integrated approach to assessing the stability, resilience and efficiency of an ecosystem to improve human and ecological welfare. In many ways its operational principles are transdisciplinary in nature (Smit et al. 1998).

Conclusion

The Ethiopian project that initially used a systems approach to diagnose problems provides further evidence that addressing complex problems of human development and environmental management requires moving from disciplinary to transdisciplinary research. Discipline-based component technology was developed and tested with farmer participation, though often separately, and assessment of impact was done mostly in bio-economic terms (taking yield and income as the norm). Over time, the need for simultaneously assessing economic, social and environmental effects of several technology interventions at plot, household and watershed/community levels was recognized. Research methods were modified accordingly to adopt an integrated and holistic approach using the agroecosystem health approach as an integrative analytical framework. This evolution also entailed a gradual shift from disciplinary to multidisciplinary to transdisciplinary research.

Acknowledgments

We gratefully acknowledge financial support at various stages of the project from the Swiss Development Cooperation, Oxfam-America, the Government of Finland, Caritas, the Government of the Netherlands, the International Development Research Center (Canada) and ILRI.

5.3.2 On the Search for Ecojumps in Technology: From Future Visions to Technology Programs

by J. Leo A. Jansen, Geert van Grootveld, Egbert van Spiegel, Philip J. Vergragt, Wilma Aarts and Conny Bakker, Sustainable Technology Development Program, Delft, The Netherlands

Recognition
The project "Search for Ecojumps" was chosen as a convincing concept for sustainable technology research and development by backcasting from needs to products and from future to present.

Organization and Description

A national research program, entitled "Sustainable Technology Development" (STD) was instituted by five Dutch ministries in January 1993 and completed in December 1997 (Jansen 1993). Industry, scientific institutes and government cooperated and invested 30 million Dutch Florins (US$15 million) in this 5-year program. About 600 technologists and other scientists participated actively and roughly 2000 individuals, including policymakers, attended congresses, meetings, and workshops.

Inspired by the World Commission on Environment and Development report, *Our Common Future* (Brundtland 1987) and in keeping with Dutch National Environmental Policy Plans, the STD program addressed a basic question – whether, how, and to what extent technology can contribute to satisfying the needs of future generations by bridging the tension between ecology and economy. The program mission was five-fold: (1) to explore and illustrate (2) together with policy makers in industry and government (3) how technology development can be shaped and organized (4) from a future orientation based on sustainability and (5) develop instruments to implement this. Researchers and policymakers were particularly interested in how options can be initiated and managed.

The Research

The program was embedded into Dutch policy-making practices in environmental policies. These practices include participatory policy making, source orientation, internalization and self-regulation of target groups within frameworks established by the government on the basis of integrated long-term planning with quantitative objectives.

Two assumptions were made:

- That the expected growth of the world population, the justified desire for welfare in developing countries, a moderate economic growth in industrialized countries and the need to reduce the environmental burden require a jump-like development of technology in fulfilling people's needs in the coming decades with increases of eco-efficiency by factors ranging from 5 to 50.
- That such development requires decades of developments that range from shaping new concepts to viable marketable products, taking into account that not a few items are concerned but the whole of technology needed to fulfill people's needs.

In constituting the program, four features were incorporated: concerted action and participation in communicative learning by doing; interaction among culture, structure and technology; backcasting from future to present and from needs to products; and iterative and interactive search.

Interaction

Eco-efficiency improvements need to be achieved for the entire range of means that fulfill people's needs, from simple items up to complex technological systems. This achievement requires intensive changes in sufficiency (a matter of culture), efficiency (a matter of structure), and effectiveness (a matter of technology):

- *culture*, legitimating the nature and volume of societal needs to be fulfilled (value sufficiency);
- *structure*, the economic and institutional organization (value efficiency) for fulfilling legitimated needs;
- *technology*, providing technical means by which needs are to be fulfilled (value effectiveness).

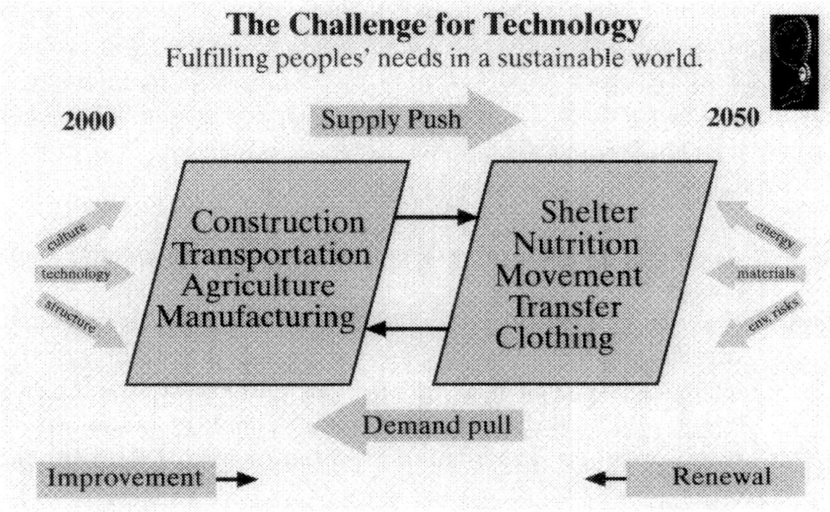

Figure 1
Backcasting from Needs to Products

Backcasting

Mindful of the Brundtland report, fulfillment of peoples needs in a sustainable future was the central issue in the STD Program. To increase technology jumps in the conceptual phases, the future (50 years ahead) was taken as the starting point for an overview and descriptions of the necessary development process. Together, these activities comprise "backcasting" (Goldemberg et al., 1985), Backcasting from needs-to-products and from future-to-present offers the possibility of identifying intermediate steps, which by themselves may be product developments. In this way, the tension between long-term societal goals and short-term (private economic) objectives can be bridged (Fig. 1).

Iterative and Interactive Search

In earlier Dutch policy research, experience was gained with iterative and interactive working on techno-economic policy making. The core of this approach is step-by-step iteration between "production" at an operational level and "evaluation" at a managerial level of subsequent guiding documents. Sample documents include project plans, problem descriptions, description of alternative solutions, criteria for choices, choice of solutions, and development plans. All iterations are interactive, involving relevant stakeholders with their interests as well as relevant scientific and technological disciplines.

Three steps were distinguished in the research:

Step 1: Analyzing domains of need and sectors to identify major challenges

Step 2: Elaborating specific needs in Illustration Processes to result in a considerable accentuation and more detailed definition of the technological challenge; to offer an outlook on the further course of development of sustainable technology and to offer a perspective on programming Research and Development (R&D) activities.

Step 3: R&D program: description and presentation of technical options in such a way that structural and cultural conditions for implementation can be recognized; the public in general can generate a fair imagination of a solution and its implications; technicians can evaluate the chances of solving remaining technological problems and uncertainties.

Transdisciplinary Operations and Results

In STD-operations, transdisciplinarity is a condition of finding practicable solutions. Transdisciplinarity encompasses not only scientific disciplines but also parties who look upon the research objective from different perspectives. Renewal of technologies may well lead to crossing the borders of known economic sectors and may require cooperation of public and private institutions, as well as governmental and non-governmental organizations. Transdisciplinary cooperation and integration involves scientific disciplines, sectoral stakeholders and institutional partners in an interactive and iterative search.

In the STD-approach the nature of transdisciplinary cooperation and integration ranges from conceptual to implementational dimensions, depending on the phase of the procedure:

- Developing a long-term vision: conceptual,
- Developing a short-term approach: operational,
- Embedding into research and planning: implementational.

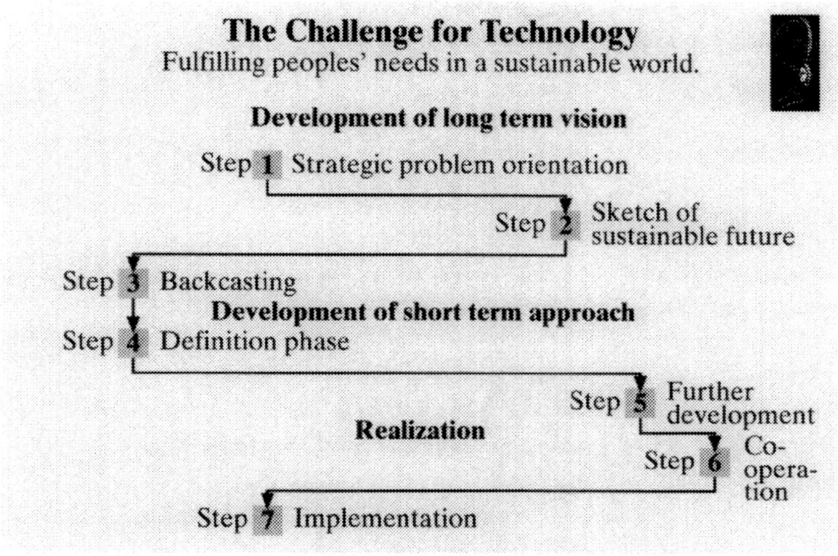

Figure 2
The STD Approach

Results

The main result of research is an approach that can be used to develop further a synthesis of ecology and economy. Weaver et al. (2000) provide a descriptive evaluation of the program in its context. An independent evaluation has been made for the Deutscher Bundestag (1999), and the program has produced a manual with recommendations on how to implement new research directions, knowledge, and technologies based on complementarity of innovations in technology, culture and structure (Jansen et al. 1998). The proposed approach is shown schematically in Figure 2.

The illustration projects resulted in a body of tangible results drawn from fifteen projects devised by interdisciplinary teams. The teams were comprised of individuals from business, centers of excellence, social organizations and government agencies. The results are new technological directions, research agendas, organizational structures, products, and business opportunities. The elaborated illustration processes belonging to different domains of human needs or economic sectors are listed in Table 1.

Table 1
Illustration Processes in the STD Research Program.

Domain	**Illustrative process**
Food supply	Novel protein foods Sustainable land use Hightech agroproduction Whole crop utilization
Transport systems	Pipeline transportation of goods Computerized processing of transport demand Hydrogen for mobile applications
Shelter	Sustainable district renewal in Rotterdam Sustainable office building
Water chain	Integrated sustainable urban / rural waterchain
Chemistry	Conversion of hydrocarbons New (organic) cells for photovoltaic solar energy Whole crop utilization Fine chemistry process technology Natural fibre-reinforced composite materials

Methodological Results

The central objective of research was to evaluate the effectiveness of the procedure for designing a desired future orientation and backcasting from there to the present, in order to construct a realistic long-term innovation path. This approach proved successful in initiating innovation trajectories and embedding them into existing research and policy institutions.

Several important lessons emerge:

- First, the results of innovation experiments result in the following indications. Shared rough estimates of future orientations among relevant stakeholders are essential elements in innovation. Investment of time in stakeholder analysis based on (bilateral) interviews is essential to creating continuous support and chances for embodiment.
- Second, operationalizing the essential interaction of "culture-structure-technology" requires paying ongoing (permanent) attention to the element of "culture." The nature of the operationalization differs by levels in the program, ranging from "future visions" to "product viability."

- Third, backcasting from needs-to-products and from future-to-present is a powerful tool for generating creative approaches to the innovation process.
- Fourth, the tension between the industrial need for economic prospects in the medium term and a necessary orientation in long-term targets for sustainability can be bridged. In fact, long-term envelope curves can be constructed "ex ante," covering and orienting technological development trajectories with intermediate targets and spin-offs.
- Fifth, the tension between the needs of broad support and of innovative creativity can be bridged.
- Sixth, networks on the interfaces of different technological disciplines and between technological and other disciplines can be set up, maintained and operationalized. Profiles for key people in the innovation process were established.

Conclusion

Within Dutch environmental policy practice, the research program "Sustainable Technology Development" was set up to define the role of technology in sustainable development. The architecture of this program was based upon several factors:

- Basics: "Cooperation among Industry, Science and Technology, Government " and "The Challenge, providing long-term future needs";
- Principles: "Interaction of Culture-Structure-Technology" and "Needs, the starting point for technology renewal";
- Methods: "Backcasting from future views and from needs" and "Iterative and interactive search."

Transdisciplinarity was an essential feature throughout the whole program. Its guiding manual showed that being able to open innovation processes leads to options for sustainable technologies.

Follow-up

A dissemination program with a budget of 5 million Dutch Florins for a period of three years (1998–2001) has been instituted to transfer resulting knowledge and know-how. This program consists of an educational part and a "learning-by-doing" part. According to a government decision in November 1998, a ten-year national initiative for searching and initiating new innovation programs was launched on 23 December 1999 (with a budget of 28 million Dutch Florins for the first four years).

The fifteen illustrative projects led to R&D programs embedded into the Dutch knowledge infrastructure. Several institutions and research centers have adopted and incorporated findings of the research in their own activities. Technologists, scientists and policymakers in business and government who participated in the program have actively spread and apply their experiences.

All in all, the Sustainable Technology Development program opened opportunities for the Dutch government, local authorities, the business world, and the science world to initiate sustainability-directed innovation processes.

5.3.3 Cross-disciplinary Knowledge as a Guide to the Study and Management of Complexity: The Case of Product Definition in the Aerospace Industry.

by Paul Jeffrey, Peter Allen, and Roger Seaton, International Ecotechnology Research Center, Cranfield University; Aileen Thomson, Warwick Manufacturing Group, Warwick University, UK

Recognition

The project "Complex System Research" was chosen as a paradigm for applying systems analysis to structure the complexity of cross-disciplinary knowledge in product design in the aerospace industry.

Introduction

Exploitation of complex systems thinking in industrial and commercial contexts is currently at a critical stage in its development, emerging from the promise of theory and rhetoric into the more critical arena of application. Opportunities for implementation of complexity concepts range from organizational structures and management paradigms, on the "soft" side, to engineering design and logistics, on the "harder" side. While a multitude of academic specialties have found intuitive meaning in the concept of complexity, we suggest that the very nature of the concept precludes any disciplinary bounding of its application. Complex evolutionary systems are by definition composed of different types of phenomena carrying out a variety of functions. Therefore, an understanding of interdisciplinary and transdisciplinary issues is a major component of complex systems thinking, in terms of both executing research on the subject and implementing subsequent prescriptive practice.

This chapter does not take its cue from an intellectual field or organizational characteristic. Its terms of reference are driven from a problem context – product definition in the aerospace industry. The research on which it is based is charged essentially with exploring how complex systems ideas might be applied to the product definition stage of aircraft design. The particular aspect of this undertaking we address here is the relationship between complex systems phenomena and interdisciplinary inquiry.

What is Complexity?

Let us begin with a brief discussion of the nature and relevance of complex systems thinking, concepts, and modeling. Over recent decades, research into the behavior of non-linear dynamics and complex systems has demonstrated they are of great importance as a new basis for understanding many types of phenomena (see Nicolis and Prigogine 1989 for an overview). However, this does not imply they provide a basis for mechanistic prediction and optimization. Rather, the natural sciences have moved beyond the mechanistic paradigm into more difficult realms of evolving and self-organizing systems, where mechanistic certainties no longer apply and emergent properties and problems occur. In such systems, qualitatively different states of organization can occur spontaneously, and a nested hierarchy of co-evolving structures is the natural outcome (Allen 1997).

Non-linear interactions between individuals or micro-elements lead to the occurrence of symmetry-breaking instabilities. These instabilities correspond to changed dimensions of description and to qualitative change in the variables and parameters relevant to understanding of what is going on. Clearly, such a picture allows us to understand the reasons why any particular model of a complex system, arrived at by making assumptions about the stability of the description and cast in terms of average behaviors and processes, will generally be inadequate and possibly short lived. If the situation under study is developing fairly rapidly, then assumptions about a fixed taxonomy and average behaviors will not be valid. It will also be necessary to look for a fuller, more evolutionary description of the system. A conceptual framework allows us to begin to understand the conditions of validity of different possible approaches. These stretch from qualitative analysis – through descriptions including learning, adaptability and evolution – to more mechanical system-dynamics models.

We should bear in mind that complex systems are more than just complicated. They undergo qualitative and seemingly unpredictable change. We can describe the significance of this difference by considering how the inordinate richness of reality is reduced to mechanistic representation in most modeling activities. The distinction rests on four assumptions that are made when representing real-world systems:

1. that one can put a boundary around the system and explain something from what is inside;
2. that one knows how to classify the parts (a qualitative analysis to choose the variables);
3. that within each category X, individuals are average (there is no micro-diversity);
4. that events occur at their average rate, defining "typical" mechanisms (no luck or local circumstance).

By adopting assumptions 1, 2, 3 and 4 plus assuming stationarity, we get Equilibria, Cycles or Chaos. With assumptions 1, 2, 3 and 4 by themselves, we get deterministic system dynamics. With only assumptions 1, 2, and 3 we get self-organizing dynamics, and with only assumptions 1 and 2 we get evolutionary complex systems.

The Significance of Cross-disciplinary Knowledge Structuring

So, what is the connection with structuring cross-disciplinary knowledge? Let us look at the previous description in more detail. It spoke of "non-linear interactions between the individuals or micro-elements," of "changed dimensions of description," of "qualitatively different states of organization," and of "qualitative change in the variables and parameters relevant to an understanding of what is going on." All of these comments suggest that more than one qualitatively different category of phenomena is interacting and requires analysis. Much of the complexity is due to interactions between incommensurate types of process or phenomena and the qualitative restructuring such interactions drive.

This assertion of the central role that cross-disciplinary knowledge has in the study of complex systems is not an artifact of the way in which complex systems thinking has been described above. The terms "complex systems" or "complexity" are often used as a general expression to describe a range of conceptual tools. If we consider some of the processes and phenomena that can be listed as characterizing complexity, additional evidence emerges for the utility of cross-disciplinary knowledge structuring as it may be understood in terms of relationships between different fields of understanding. Let us consider some examples.

Co-Evolution in ecological terms can be seen as the process through which different species (flora and fauna) adapt and change through interaction with each other and with other environmental moderators.

Each one is partially dependent on others for its particular evolutionary trajectory. This symbiosis of development has clear parallels in, for instance, industrial development. Many industrial sectors are dependent on the activities of other sectors for their survival, and they are influenced by political and economic trends. Consequently, any attempt to understand co-evolution requires involvement of a range of disciplinary contributions and effective structuring of the various contributions.

Another common complex-system concept is *Fitness Landscape*, which represents the fitness measure of a problem. Most real problems are not uni-dimensional, and the majority of fitness landscape peaks and troughs can only be effectively measured through compound or multi-dimensional (cross-disciplinary) values. Perhaps the most intriguing possible application for cross-disciplinary structuring is in the area of *Self-Organization*, which refers to spontaneous emergence of macroscopic non-equilibrium organized structures due to collective interactions among a large assemblage of simple microscopic objects. Similar cross-disciplinary opportunities can be found in the areas of *Emergence* (appearance of higher-level properties and behaviors of a system that are properties of the whole not possessed by any individual parts making up that whole) and *Resilience* (ability of a system to suffer degradation without altering its viability).

By applying more than one method to analysis of a problem, interdisciplinary analysis formally introduces the complexity of the real world into a decision-making process. Two features of interdisciplinary analysis serve to emphasize this point. First, the search for optimal solutions has been a characteristic of problem analysis methodologies since the emergence of the empirical scientific tradition in the seventeenth century. Optimal solutions simply do not exist for a whole range of problem types (or if they do, the definition of "optimal" is so broad as to warrant use of an alternative term). An interdisciplinary approach, by its very nature, shifts the focus away from ideal solutions towards alternative criteria such as the level of consensus which different options attract, their achievability, and how they contribute towards overall system sustainability. Second, involvement in interdisciplinary analysis processes engenders an appreciation of alternative perspectives. This serves to promote an investigative/exploratory element into analysis of decision issues and encourages development of response options rather than problem solutions.

Finally, we turn to a brief example of how cross-disciplinary knowledge structuring is being used to explore issues of complexity.

Product Definition in the Aerospace Industry

The aerospace industry (in terms of total aircraft design and manufacture) is a dynamic, high-tech, engineering-based sector. It operates within a unique environment characterized by extended project horizons (typically eight to fifteen years) and a strong emphasis on safety and maintenance of credibility. Product definition within this sector comprises a set of processes through which a product "becomes"' and attains final form. Very often, these processes do not stop when a formal product-design stage has been completed. They continue through various production and development activities. Complexity arises through the connectivity and inter-relatedness of the system's constituent elements.

The way these interrelationships arise in the product-definition phase of product development is of particular interest. The challenge is to understand how a particular aircraft configuration emerges from the set of all possible configurations. What does the total design space look like, and are some regions of it being ignored? The relationships inherent in the process of product definition may be seen in terms of adaptive evolution. However, adapting entities confront conflicting constraints, both in internal organization and in interactions with their environments. These conflicting constraints typically imply that finding the "optimal solution" is very difficult. Yet, it also means that many alternative locally optimal compromise solutions may exist.

Furthermore, the consequence of attempting to optimize in systems with increasingly conflicting constraints among components brings about what Kauffman (1993) calls a "complexity catastrophe." As complexity increases, the heights of accessible peaks recede towards the mean fitness. The onset of the catastrophe traps entities on a local optimum, thus limiting selection. This is clearly important as it applies to the product-definition process needing to cope with an increase of conflicting constraints. Our central interest here is understanding how the design space (in engineering terms) or opportunity space (in decision-making terms) evolves and narrows down to become the final product.

Our initial inquiries were in the form of an interview-based problem-diagnosis exercise. It revealed a wide range of criteria that have an impact on the shape of the design envelope at any point in time. Economic, regulatory, financial, contractual, project time-scale, environmental, and competition issues are just some of the non-engineering considerations influencing design progress. And, this is just for the aircraft as a whole. When you start breaking down various components and considering their sub-components in terms of design/attribute spaces (each one

Table 1
Structuring of Cross-disciplinary Design Contributions to Form an Interview Agenda

Design envelope evolution themes ↓	Academic / Professional Discipline										
	IT support	Simulation environments	Accounting	Legal department	Project management	Strategic requirement concept development	Production / Manufacturing	Sales & Marketing	Aeronautics	Thermal design	Mechanical design
Whenever a projection or expectation is made into the future, how is this made ?				X	X	X			X	X	X
What is the knowledge base and model used to project forward?	X	X		X		X	X		X	X	X
What factors increase / decrease the possible option space ?			X	X		X	X	X	X	X	X
What are the criteria of evaluation and where do they come from ?			X		X	X	X		X	X	X
At what cost are options held open ?			X	X			X	X			

being managed by a different company), the problem becomes both complicated and complex. An exploration of design-envelope evolution clearly requires understanding information and data from several intellectual fields. Accordingly, we have structured the various disciplinary contributions to the problem set in the manner shown in Table 1. The core research issue has been split into contributing themes (derived from the diagnosis exercise). Each theme, in turn, is associated with a set of disciplinary knowledge bases.

Conclusion

Our research is still on-going. Therefore, it would be unwise to draw any sweeping conclusions about the robustness of our ideas or the utility of our approach. However, we are encouraged by the way in which the inquiry structure represented in Table 1 has prompted interviewees to consider the interface between their own experiences and tasks, and those of their co-workers. The content of our data set (comprised of transcribed interviews) is thereby distinctive. A question remains as to whether we can distinguish cross-disciplinary phenomena from the dataset. Discussions with colleagues during the International Transdisciplinarity Conference suggest that the type of cross-disciplinary phenomena we are seeking to identify may require interpretation by respondents or researchers with genuinely transdisciplinary skills and knowledge. If this is indeed the case, then the question of how one would go about characterizing such skills is clearly an issue of some significance.

Acknowledgments

The authors acknowledge the financial assistance of the UK Engineering & Physical Sciences Research Council, which supported this research under Grants GR/M23649 and GR/M24226. The case study information on which the paper is based came primarily from our industrial collaborators at BAE Systems, Rolls Royce and HS Marston.

5.3.4 A Modest Success Story: Linkki 2 Research Program on Energy Conservation Decisions and Behavior

by Pirkko Kasanen, TTS Institute, Helsinki, Finland

Recognition

The project "A Modest Success Story: LINKKI 2" was chosen for clearly pointing out the crucial role of elements for successful transdisciplinary research, especially participatory definition of goals and questions.

Introduction

Transdisciplinary research is often carried out in answer to societal challenges involving the environment. Climate change is a major global challenge for both research and policy. As a member of the EU, Finland has a climate policy to reduce greenhouse gas emissions. Finland's target is to limit CO_2 emissions by 2010 to 1990 levels. The energy conservation program is a part of this policy.

From 1991 to 1995, the Academy of Finland funded the first major effort to encourage social science research on environmental issues in a Research Program on Sustainable Development. From 1990 to 1995, the Finnish Research Program on Climate Change, SILMU, was funded for 77.9 million Finn Marks. It was an interdisciplinary program combining inputs of natural and social sciences. From 1999 to 2002, the Finnish Global Change Research Program (FIGARE) has been running at a budget of 40 million Finn Marks. Its projects represent a wide variety of disciplines.

In these programs, gradually increasing attempts have been made to integrate various disciplinary research approaches. In the Sustainable Development Program, no integration of individual projects was considered possible or seriously attempted. In SILMU, an integration project was set up at a late stage to answer questions of policy relevance. It did not have a chance to proceed, though, in an iterative manner between research groups. FIGARE has a full-time coordinator who is planning to keep close contacts with administrators needing results of research in their work. In FIGARE, research consortia representing a variety of relevant research fields were also encouraged.

On a much smaller scale, the Ministry of Trade and Industry in the Department of Energy finances LINKKI 2, a five-year research program (1997–2000) on energy conservation decisions and behavior. More than half of the program's annual budget of between 2 and 3 million Finn Marks comes from the Department of Energy. LINKKI 2 investigates the functions that connect or separate technical possibilities for energy conservation and users of the technology (as consumers or in other roles). It aims to combine the means available to technology and economic research with social and behavioral sciences.

Due to the relatively small budget and pressing character of the energy conservation challenge, the Department of Energy expects approaches and results that can be used. Disciplinary purity is by no means a consideration, and more or less happy matches between approaches are made in pursuit of practical results. In this chapter, I discuss the ways in which the user orientation and matching of relevant disciplinary approaches are arranged in LINKKI 2. I outline the structure, content and functions of the program. Then, I review some previous and ongoing studies to identify how interdisciplinarity and user orientation have been arranged and what particular arrangements seem to have been successful. These findings illuminate the characteristics of success in such a project.

LINKKI 2 on a Learning Curve

LINKKI 2 is based on two experiences: in 1990–1992 the Department of Energy funded a research program on Energy and the Consumer and, in 1993–1995, a LINKKI Research Program on Consumer Habits and Energy Conservation. This sequence reflects changing emphases in both energy conservation policies and research. The Department of Energy's involvement in outlining goals and contents has increased from one program to the next, and even during LINKKI 2 itself, where user orientation emerged as a key concept.

Before the first call for applications in LINKKI 2, the Department of Energy drafted a description of intended subject areas and their main objectives. A steering committee of ten people was set up, consisting of representatives of intended users of results. The representatives were the Energy Department, MOTIVA Energy Information Center, Ministry of the Environment, Ministry of Transport and Communications, The Association of Finnish Local Authorities, The Finnish Electricity Association,

The Finnish Real Estate Federation, and, as potential co-funders, The Academy of Finland and TEKES.

The first call for applications, outlining rough subject areas, received a mixed response. In subsequent application rounds, description of objectives has been increasingly fine tuned according to the most essential conservation policy needs by the Energy Department. The actual research is carried out by individuals or groups. Internal communication of the program and the input of users is organized by means of follow-up groups, in which users meet with researchers of one or several related projects. In the latter case, researchers also get to exchange ideas and experiences with each other.

Results are publicized in a program series and abstracts are on the Internet. Cooperation with users, which has been an important way of disseminating results, has sometimes resulted in more focused brochures, articles, and seminar presentations.

So far, in three application rounds, there have been twenty projects in the following subject areas:

- Households and residential buildings (4 projects)
- Municipalities, companies, organizations (9 projects)
- Follow-up and efficiency of measures (4 projects)
- Adoption of new technology and new measures (3 projects)

Managing Inter- and Transdisciplinarity

The cross-cutting project, "Theories for Praxis, Praxis for Theory," which is run by the German Urban Ecology program, investigates cooperation of natural and social scientists within substantive projects of the program. While the system theoretical approaches and conditions of participation of different disciplines on different levels of analysis are an essential part of the study, these issues are beyond the scope of the much smaller LINKKI 2 program. Hence, I focus here on practical experiences of integration.

Three basic strategies have been identified that foster coordination and cooperation within an interdisciplinary group (Hollaender and Friedrichs 1997; <http://www.itas.fzk.de/deu/TADN/TADN397/schwer.htm#schwer7>).

(1) Having a very clear goal from the beginning is crucial, one that representatives of different disciplines can agree on.

(2) It is also essential to leave in the background differences that are, nevertheless, recognized.
(3) Some minimum of conceptual definition is needed as well.

Three strategies operate in LINKKI 2. Each project has a clear goal that overrides possible disciplinary interests. Groups build on extensive networking and information dissemination. Related projects are organized into networks within which information dissemination occurs. Researchers also have a common theoretical approach, such as a system theoretical approach. The most demanding part is the need to adjust disciplinary contributions to fit in a common model. In LINKKI 2 there is no such common theoretical approach.

The DACH project (*Defila et al.*) which studies control procedures for interdisciplinary and transdisciplinary environments in German, Austrian, and Swiss cooperation, has sent questionnaires to 600 researchers. The questions refer to relatively large projects or modules. A variety of alternative answers is offered for each question. Keeping in mind the small scale of LINKKI 2, I have selected the most relevant questions that represent themes in LINKKI 2.

- What were the important points in developing common goals and questions?
- What were the important conditions in integration of a module?
- How was a common language developed in the module?
- What communication tools were used in internal exchange of information, results, etc.
- How were results distributed outside the research group?
- What were the important factors in including user needs successfully in the module?

Two background questions are also noteworthy:

- How is research oriented as a rule (on a scale from user orientation to basic research)?
- What were the roles of different disciplines and institutions in the module?

Eleven projects of LINKKI 2 are now complete. "Success" is measured subjectively by asking what users or intended users have to say about using the results. Yet, the evaluations are quite clear. The development of goals and questions was very praxis-oriented from the beginning.

However, the Department of Energy shaped ideas during the process. Therefore, projects starting in the second round or later are more clearly user-oriented than those in the first application round. All later ones can be regarded as successful, but only one of the earlier ones was based on previous practical work with one of the users.

On the level of individual projects, there is no need for integration or developing a common language, since the projects are carried out by small groups of researchers with similar backgrounds. Some groups within projects work around common themes, however. Meetings, e-mail and printed drafts are typical means of communication. While this work is essential for the program, it does not differentiate between successful or less successful projects. Successes have been both individual and in subsets of larger groups. The task of the part-time coordinator is to see to this information exchange between projects. In addition to project meetings, the program organizes annual seminars for researchers and users.

As for activities that distribute results outside research groups, some forms of user involvement are prominent in successful projects. User needs are included by the steering group, but, as noted earlier, this is only related to success when clearly specified. In successful projects, users participated in developing "products," such as guidelines, background for information material, and energy-efficiency data-collection tables. Some of this development takes place after the project, based on results, though a continuum is apparent. Financial involvement of specific users has not been substantial, apart from funding by the Energy Department.

Regarding background variables, all of the researchers do mostly user-oriented research, though some are at the same time involved in the academic world through Ph.D studies. The projects are rather small, not allowing major involvement in the theme. To maintain high quality, the institute must on the whole be deeply involved in the theme and have previous experience. An individual researcher must have an ongoing, deeper interest in the area through Ph.D studies, thereby giving a solid theoretical framework. Successes occur in both instances.

The program includes social sciences and economics, humanities, and technology. By institutional background, universities, public and private research institutes and consulting companies are represented. Successes exhibit the same mixture of backgrounds as the program in general.

Recipe for Success

The involvement of users from the beginning is an iterative process of establishing questions. An idea is needed: how to develop energy efficiency "products" based on research. This process demands resources from users, not necessarily in the form of financial involvement in the research itself but always in terms of time and commitment.

From the point of view of researchers, the practical goal makes it possible to contribute what they can, rather than watch disciplinary boundaries. New, often inspiring connections are made that can be useful to researchers in the longer run as well. It is not a question of giving up one's integrity, either. On the contrary, user orientation has to be based on being able to trust the academic quality and credentials of researchers. The researchers, in turn, need a solid academic or institutional basis in order to be able to afford participating in these relatively small projects.

Extra resources are also needed to process results into forms that can easily be distributed to those who are interested. Traditional research reports are not very good at this, but researchers cannot easily turn into public relations personnel either. A coordinator is needed to organize the framework for user and research communication, and sometimes as an interpreter between approaches.

5.3.5 Transdisciplinarity in Planning of Sustainable Urban Revitalization

by Paulius Kulikauskas, Niels Andersen, Freddy Avnby, Lykke Leonardsen, Ole Damsgaard and Andreas Schubert, Byfornyelse Danmark (Danish Urban Renewal Company), Copenhagen, Denmark

Recognition
The project "Innovative Urban Planning and Management", which was partially financed by the EU European Regional Development Fund and was chosen for its lucid analysis of stakeholder participation and for overcoming obstacles in cross-sectoral urban planning.

Background

Many cities and countries are facing a complex problem of decaying urban neighborhoods. Dilapidating buildings and infrastructure, growing unemployment, and the decline of a local economic basis contribute to a dull environment, discouraging residents and businesses. Adversities interinduce each other, increasing the resignation of people in these neighborhoods. Even the fattest public purse seems too shallow to remedy the lasting consequences of such degradation.

When the economy starts growing, many expect that general growth will help solve the problem of decaying neighborhoods. However, the contrary often happens – the beneficiaries forsake deprived neighborhoods, increasing segregation further as their place is taken by those who are even more socially depressed. Often, efforts of authorities to redirect public spending bring about little structural change. Improved public areas and housing soon decay, and expensive achievements deteriorate quickly after public spending is curtailed.

Such experience has led public authorities in many European countries to follow two trends: reviewing public spending on isolated physical renewal and reactive support, while looking for long-term results of financial injections aimed at sustaining neighborhoods by uncovering endogenous resources. Two related questions arise:

- How can we ensure sustainable revitalization in disadvantaged neighborhoods?
- How can we develop new planning processes to satisfy the complex needs of neighborhoods?

Various public and private institutions strive to answer these questions. They have developed means of meeting community needs by improving living conditions and balancing the investment climate. They have also secured cohesion in the settlement structure by encouraging disadvantaged communities to undertake responsibility for development processes, opening opportunities for stakeholders to take part in planning and addressing a wide spectrum of complex interdependent problems, ranging from poverty to lost cultural identity.

Participation and Transdisciplinarity – Two Sides of a Coin

Experience in Denmark and abroad, plus continuing monitoring of international research, indicates there are two key components in successful planning and management of urban revitalization: participation of stakeholders in all stages of planning, decision making and implementation; and application of cross-sectoral approaches to planning sustainable revitalization. For our purposes – working in integrated urban revitalization - we limit ourselves to exchanging empirical experience and defining issues related to how disciplinarity affects planning and governance of urban revitalization, followed by elaboration of a transdisciplinary methodology (Fig. 1).

Stakeholders' interest in revitalizing an urban neighborhood, their visions and objectives, may be transdisciplinary in nature. However, when starting to work with them, things often go the usual disciplinary way. Experience in Kongens Enghave, a neighborhood in southwest Copenhagen, is a case in point. This experience is linked to an EU project aimed at networking several communities around the Baltic Sea. The broader project has several goals:

- striving to develop means of meeting community needs in order to improve living conditions,
- balancing investment climates and securing cohesion in settlement structures by encouraging disadvantaged communities in decaying urban areas to undertake responsibility for development processes,
- opening new opportunities for stakeholders to take part in planning,

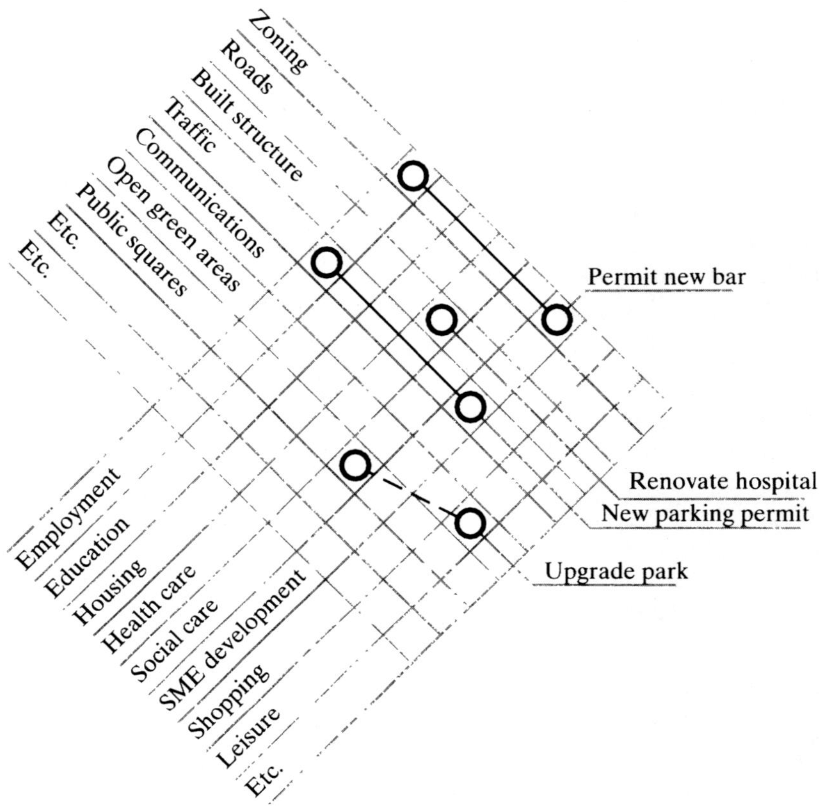

Figure 1
Cross-Sectoral Planning

- addressing a wide spectrum of complex interdependent problems, ranging from poverty to lost cultural identity.

As a result of the combined experiences of pilot projects, a set of guidelines – a "cookbook" – will be produced, presenting approaches and instruments for a new way of thinking about how to cope with related problems. The lessons gained from this experience will be further developed through cooperation in a transnational network, providing educational tools that enable all partners in the planning process to contribute to the search for a solution. (For further information, see <www.ensure.org/iupm>.)

The immediate research project in Kongens Enghave enjoys both an experimental locally-elected council and a government subsidy for urban revitalization under the Danish Urban Regeneration Experiment. When residents became involved in the planning process, they chose to form sectoral working groups, such as physical problems, housing, culture, employment, and social issues.

Why would they do that, if their initial interest was not disciplinary? The reason may be that transdisciplinarity is a complex phenomenon, and there is little experience in tackling relevant approaches. People have a natural inclination to reduce complexity by dissecting the problem into sectors. (Even in the Transdisciplinarity conference Mutual Learning Sessions were somewhat sectorally divided.) Two additional questions arise:

- How can the natural oneness of stakeholders' comprehension be retained in elaborating a revitalization program?
- How can one employ an transdisciplinary approach in developing programs?

A good metaphor, produced by the Kongens Enghave program manager, refers to sectoral and disciplinary issues as boxes. Coordination is the simple act of moving boxes next to each other. A transdisciplinary approach empties the boxes into one big pile and shuffles their contents. The trick is to repeat the cycle as often as necessary.

Because the planning process is directional, projecting it on a timeline vector "X" and laying extremities of this cycle on the perpendicular "Y" depicts process as an irregular, asymmetrical helix (Fig. 2). Moreover, it is multiple-thread. Helices would intertwine in a complex contrapuntal movement, though this parallel is difficult to grasp. Perhaps it is enough to say that a transdisciplinary approach is not a panacea, something that has to replace sectoral public governance. Rather, it is another level of approach that must be adopted in conjunction, not as replacement of, disciplinary and sectoral aspects.

The method may also be described as resolving cross-sectorally defined problems sectorally by setting goals. That would have two effects: tackling reduction of complexity in recognizing the problem and serving stakeholders' interests. Expressed in lay language – "Know where you want to arrive at, and keep your sight on the target while minding your steps."

Complexity should not be feared in the participation of stakeholders. As Americans say: "Tell me, and I'll forget. Show me, and I may not remember. Involve me, and I'll understand." In a planning process based on transdisciplinarity and participation, the starting phase is reaching

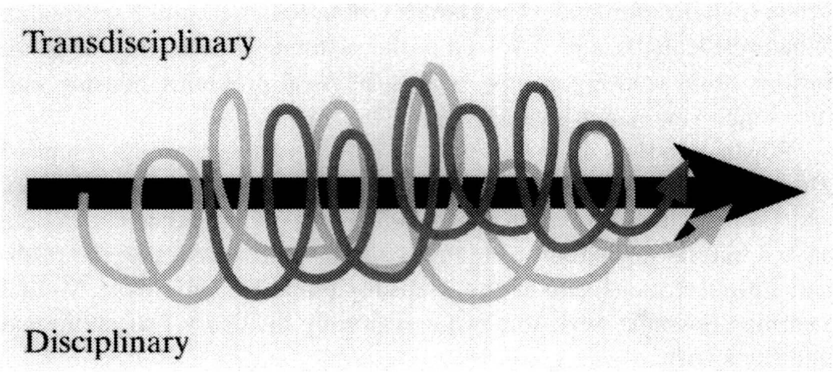

Figure 2
The Planning Process as a Helix

common understanding of the nature of the problem with deprived and regressing neighborhoods. On the basis of such understanding, stakeholders may be helped to generate a shared vision, translated by a sound strategy, into a set of principles and goals. Such a vision must contain something for everyone. Since participation requires dedication of resources, it must offer some benefit (Fig. 3).

Additionally, the vision may evolve and eventually be modified. At the other end of the planning process, partnerships may be formed to implement components of the revitalization program and carry onward. These partnerships must have well-defined objectives and a limited number of interested and dedicated partners. However, by no means do they have to be disciplinary. They must serve the interests of the partners.

The two framing phases provide a scale for participation of the widest and narrowest scopes. Achieving sustainability of revitalization in urban neighborhoods requires full-fledged participation of stakeholders. Having them understand the nature of the problem, and sharing visions and objectives leading to sustainability, requires a transdisciplinary approach. This maxim unites the two sides of the coin – participation and transdisciplinarity.

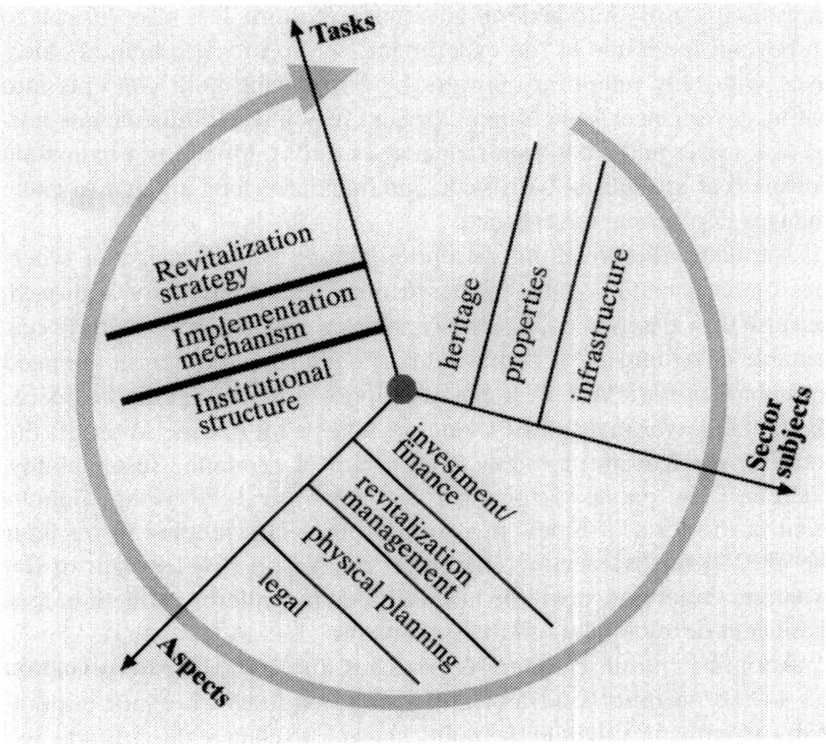

Figure 3
Combining Sector Subjects, Aspects and Tasks

Fashion in Public Governance vs Commitment to Making Transdisciplinary Participatory Planning an Inseparable Part of Governance Culture

Participation of stakeholders and transdisciplinarity are becoming fashionable subjects among public authorities. Most are quite eager to engage into experiments and demonstration and pilot projects named after transdisciplinarity, sustainability, and participation. This interest is fueled, partly, by disappointment in results of traditional urban renewal. On the other hand, there have been many policy statements on that account in multiple international forums, with repercussions finding their way into national and municipal levels of decision-making.

Things become different when one tries to set an exit strategy for a pilot project, by attempting to integrate concepts of participation and transdisciplinarity into general governance culture. It is also difficult to go beyond the frame of "an experiment" set up to attain limited objectives, with only temporary powers, by introducing both concepts into public governance. From Dublin, Ireland to Vilnius, Lithuania, integrated task forces and cross-sectoral agencies find it difficult to accumulate powers that are substantially wide and mandated long enough to make changes in planning governance.

Another widespread misconception among politicians is that synergies of participatory planning based on a transdisciplinary approach, because they uncover endogenous resources and attempts to attain sustainable development of such neighborhoods, should diminish the need for public support. Moreover, the assumption is made that market forces should take over investment sooner. It takes a lot of time to set off the complex mechanism providing even a hint of economic sustainability. Therefore, the public sector must also demonstrate a lasting commitment in the form of timely provision of necessary funding throughout the entire process, reacting quickly to changes in development of the program. Otherwise, it will be incapacitated by underinvestment before significant development of interest appears.

Adopting transdisciplinary approaches and delegating responsibilities to the neighborhood level, obviously, requires a broader outlook than currently prevalent at the political level. To change this attitude, criteria of success are needed. When a sectorally-arranged public governance system engages the process of planning, an official, for instance a traffic officer, measures success by how much positive visibility a plan and its implementation has provided for, not only for the official but immediate superiors. By setting transdisciplinary goals, one must also fight for redefinition of criteria. Otherwise, the sectorally-divided governance system has little chance to respond and to facilitate properly. This can often be achieved by delegating more powers to neighborhoods. Kongens Enghave is the only area under the Danish Urban Regeneration Experiment and simultaneously has a locally-elected council (again, a limited "experiment"). In contrast to other regeneration areas, these criteria have changed over several years, becoming more holistic and locally oriented.

5.3.6 The Role of Participatory Technology Assessment in Policy-Making

by Michael Nentwich, Institute of Technology Assessment, Vienna, Austria and Danielle Bütschi, Center for Technology Assessment, Berne, Switzerland

Recognition

The project "Political Role of Participatory TA," was chosen for its illuminating answers to the question "To which ends do we enter in participatory forms of research with the public and what is important for success?"

Introduction

This chapter starts with a short overview of an international comparative project on participatory forms of carrying out technology assessment (pTA) across Europe. It then summarizes the project itself, focusing on a comparative analysis of political roles of pTA. The concluding section addresses the political potential of participatory methods.

In essence, Technology Assessment (TA) has a strong political dimension. When the American Congress developed TA in the 1970s, it envisioned a political instrument that would give to its members access to independent, objective, and competent information on scientific and technological issues. Members of Congress would thus be in a better position to appreciate legislative projects and be able to base their political actions on more viable alternatives.

As the concept of TA evolved further over the years, particularly in Europe, several developments occurred:

- First, the audience of TA studies was not composed only of legislators but increasingly included the bureaucracy and other levels of government.
- Second, while the American model was based on a rather scientific approach (involving stakeholders only afterwards), European TA always struggled with how to integrate interests and values in the assessment.

One strand of European TA, mainly originating in Denmark, attempts to solve the problem of how to make values and interests fruitful by organizing participatory procedures. With this "participatory turn," the political dimension of TA was reinforced. It is no longer strictly an academic activity whose outcomes are to be communicated to and used by policy-makers, but a political activity itself. Integrating various actors is eminently political, because questions of power, influence and responsibility intervene.

The politicization of TA activities by integrating participatory elements originated in a recognition that the state is under pressure. New developments in science and technology put public authorities under stress. They are faced with uncertainty about the consequences of these developments and the plurality of values and interests that surround them. In this sense, development of pTA arrangements is a response to the legitimacy crisis of the state. Furthermore, our other theoretical lens – inequality – highlights the possible political contribution of pTA, especially for taking into account the plurality of views and values present in society and giving them a voice.

The EUROpTA project

The EUROpTA project was a common enterprise among six groups of TA researchers and practitioners across Europe: the Danish Board of Technology, the Dutch Rathenau Institute, the German Bureau for Technology Assessment, the TA Center of the Swiss Science and Technology Council, the University of Westminster, and the Austrian Institute of Technology Assessment. The group carried out a systematic survey of participatory forms of TA, based on a common theoretical and analytical framework. We produced 16 case studies and made an in-depth comparison with a view to five aspects:

1) introduction of pTA in new situations
2) the functional interrelationship among the objective of a participatory arrangement, the issue treated in the arrangement and the method chosen;
3) management of participatory arrangements;
4) the effects of pTA on public debate and science and technology policy- and decision-making;
5) the political role played by participatory arrangements.

Political Roles of pTA Arrangements

The major question we want to consider in this chapter is "How do pTA arrangements perform in the policy process?" Put another way, "What exactly is their answer to the legitimacy crisis of modern states?" Aimed at integrating analysis of scientific and technological developments into societal debate, participatory processes face an ambitious task. Their role within the policy-making process is also rather complex. To begin with, as a new instrument pTA arrangements still must prove worth the effort. Even then, the place they might occupy is far from obvious. Participatory TA is an addition to already highly complex political procedures and institutions that, furthermore, differ from country to country.

Consequently, pTA has to construct its role in each political system. This becomes evident from case studies in the EUROpTA project. Most initiators wanted pTA arrangements to influence the process of policy-making in some way, but with different goals and perspectives. However, the likelihood of a pTA arrangement having any political influence depends not only on the aims of the initiators but also the type of arrangement (societal and institutional contexts).

In looking at our case studies from a comparative perspective, we analyzed both the types of roles in policy-making that can be assigned to a pTA arrangement and the factors that influenced whether or not pTA will succeed in having any political function. We proceeded in three steps.

The first step was to establish an inventory of political roles that might be played by pTA arrangements (Tab. 1). Note that in items (2) to (6), the list reflects the particular stage of policy development in which the issue at stake is placed. Beginning with putting the issue on the agenda, two phases of the policy definition phase are covered (exploring objectives and filtering policy alternatives). Other possible stages where pTA may potentially play a role are cases in which policy-making is blocked or in the implementation and/or evaluation phase.

The second step was looking at practice and trying to assess the political performance of the arrangements. We concluded that the arrangements actually played a political role in some sense in all of our case studies. Even arrangements without specific political aims led to some discussion in the political arena. Note however, that political performance, in the sense of attaining envisaged political roles, was rather weak in the majority of cases. Moreover, it is difficult to assess the impact and hence the performance of an arrangement, for two reasons. First, we only seldom have in-depth impact studies and, second, it is often impossible

Table 1
Possible Political Roles of Participatory TA Arrangements

Possible political roles		**Description**
(1) Indirect political roles	a. promoting communication between science and the public	public understanding of science
	b. stimulating public debate	allowing an open dialogue between experts and non-experts and spreading information
	c. awareness building	attaining political or societal goals indirectly via raising awareness among those concerned with implementation
	d. raising sensitivity for method	changing perceptions about the forms of political discourse
(2) Agenda-setter		identifying all aspects of an issue and putting them on the political agenda
(3) Exploration of objectives		clarifying the different preferences and values as well as developing proposals for normative judgments
(4) Filter of policy alternatives		offering advice on the alternative to choose
(5) "Blockade-runner"		contributing to the management of a political conflict or stalemate
(6) Implementation and evaluation of policies		testing whether the ideas of the public are in line with the ideas of the politicians; evaluating a policy after implementation

to distinguish between effects triggered by the arrangement or by other concurrent events.

In the third step, we discussed intervening factors influencing the actual political role as well as the relationship between the following "success factors" and the political aims of pTA arrangements. This led to the following list (Tab. 2):

Conclusion

The debate about pTA attributes the role of helping decision-makers arrive at a decision only to participation. Analysis of case studies in the

Table 2
Success Factors Influencing the Political Role of pTA Arrangements

Societal context	• Good timing with public controversy • Good timing with *de facto* policy-making • Political relevance of the topic • Political culture open for (informal) participation
Institutional context	• Link to the political sphere • Credibility and reputation of the institution
Properties of the arrangement	• Precise definition of the political goals • Fairness of the process as perceived by the political observers • Product of the arrangement aiming at practical implementation • Involvement of political actors in the process

EUROpTA project showed that the political role of pTA is far more complex and is related to the entire policy-making process. Moreover, whereas most pTA arrangements seek a direct political role, some try to intervene in the policy-making process in a more subtle manner, for instance by stimulating public debate on the issue or by raising sensitivity for public participation. Yet, while discussion of the case studies revealed that pTA has an inherent political dimension that can be recognized in goals set by organizers, their actual political role often falls short of expectations.

Many intervening factors influence the political performance of a pTA arrangement. The "political success" of an arrangement depends not only on one or two favorable factors but a particular combination of factors. This is true, for instance, of all consensus conference-like arrangements outside of Denmark. Although the procedures are well tested in the Danish context and proved to be working well, so far various circumstances and factors have led to rather poor results (in political terms) in other countries. Nevertheless, our study has revealed some factors that must necessarily be met for every pTA arrangement to be successful, regardless of the particular aim. Credibility of the institution and quality of the process are key factors. In addition, other factors are important, though their weight depends on the political role that an arrangement aims at.

From these findings, we can make two general recommendations with respect to the political dimension of pTA.

- First, when practitioners envisage the possibility of setting up a pTA arrangement, they must be conscious that, in some way or another, they will act within the given policy system. This may be intended or unintended, and the intervention can be strong or minor. As a consequence, and independent of considerations related to the institutional setting and properties of an arrangement, it is important that prior to starting any pTA the actual political situation be carefully considered. Insights into the timing and the political relevance of the issue will be gained. The aims of the pTA should then be adjusted to this political situation. This will make it much easier to gain influence on the policy-making process. Moreover, since gaining influence is a matter of communication, implementers should communicate the results of the arrangement.

- Second, the audience of pTA arrangement – politicians – should be considered carefully. Often, pTA is viewed as an instrument for giving advice to politicians on which decision is the right one. This chapter indicates that the influence of pTA on the policy-making process is of a much more subtle nature. It can, for example, contribute to putting an issue on the agenda, sketching the direction a specific policy should follow, or overcoming blockades. Moreover, typically pTA is expected to create something new or rescue politicians from the position of a "non-decision." However, the cases we observed reveal that in many cases pTA is, instead, a catalyst. Minority proposals are also presented as viable solutions and have a chance of being accepted by the majority. In still other cases, pTA can bring new ideas that will develop in time and even generate further new ideas. In this respect, the influence of pTA on the policy-making process is of a very special nature. Finally, when assessing the role of pTA on policy-making, we must not forget that the actors intended to take up results of the pTA do not always agree with its outcomes. pTA is always part of the political game in which power is at stake.

Acknowledgments

This chapter largely draws on a paper the authors wrote as part of the EUROpTA project. It was first presented in The Hague on 4-5 October 1999 at the 2nd International EUROpTA workshop, then became part of the final report to the European Commission in January 2000. An extended version will appear in 2000 in a book published by University of Westminster Press, edited by Simon Joss and Sergio Bellucci. The EUROpTA project homepage maintains up-to-date information and full texts of papers and case studies: <http://www.tekno.dk/europta/>.

5.3.7 The Potential of a Research-Action-Capacity Building Approach for Effective Management of Urban Environmental Problems

by Kaspar Wyss, Swiss Tropical Institute, Basel, Switzerland; N'Diekhor Yémadji, Centre de Support en Santé Internationale and Department of Geography, University of N'Djaména, Chad; Abdias Nodjiadjim, Centre de Support en Santé Internationale, N'Djaména, Chad; Mamadou N'Diaye and Enda Graf, Dakar, Senegal

Recognition
The project "Urban Environmental Management" was chosen for effectively strengthening community initiatives in a participatory action-research approach in a North-South research partnership for the management of urban environmental issues in West Africa.

How Can Effective Management of the Urban Environment Be Achieved?

When concepts are developed for sustainable management of the urban environment, technical approaches must be in synergy with the cultural, social and economic realities of the people themselves. Many factors are relevant to any attempt to improve living conditions in the city. They include social organization, traditions, local leadership and conflicts, collaboration and communication, interaction and networking, community initiatives, and control of access to different resources. It is important to identify approaches to planning that will encourage consideration of these factors and their inclusion in concepts for urban management. Additionally, ways of developing synergy between different categories of actors and institutions need to be examined.

Since 1994, the Swiss Tropical Institute (STI) has been testing the potential of a Research-Action-Capacity Building (RAC) approach. This project, named "Management of a Deprived Urban Area by its Inhabitants," was framed within Module 7 of the Swiss Priority Program Environment. The main location was N'Djaména (Chad), with parallel activities in Dakar (Senegal) and Ouagadougou (Burkina

Faso). Work has been done in three major areas: a) water and waste management in low-income urban areas; b) social mobilization of marginalized people, especially "street children," and c) health promotion through marketing of impregnated mosquito nets for malaria control.

The project has three objectives:

- to empower grass-root initiatives for better urban environmental management, through a process that includes collaboration through participation, and acquisition of knowledge by popular and institutional actors;
- to introduce and maintain the practice of collaboration among researchers and popular and institutional actors (such as municipalities, NGO's, multilateral agencies);
- to promote regular networking and sharing of experiences among different cities in the African region and between popular and institutional actors.

The Action-Research Approach in a Transdisciplinary Setting

South-South and North-South research partnerships, and collaboration at various levels, are vital to increasing and strengthening relationships and exchange of expertise in urban environmental management. Therefore, a transdisciplinary framework was an essential part of the project. Ideas, concepts and activities were developed by and shared among local residents (associations), institutional and professional actors in the city, concerned municipalities, NGO's, and other institutions located both in the South (Enda Graf, CREPA, EIER) and in the North (Swiss Tropical Institute, Universities of Avignon and Strasbourg).

The selected approach, Research-Action-Capacity Building, can be described as a systematic inquiry undertaken by participants in order to improve their own practices and to deepen their understanding of these practices and of situations in which they are carried out (N'Diaye 1999; Yémadji et al. 1999; Wyss 1999). In contrast to a top-down approach, RAC is an iterative process that includes the people concerned directly in planning, execution, and ongoing evaluation of activities (Fig. 1). "Capitalization" of experience at all levels of this cycle is crucial. The following examples illustrate this approach.

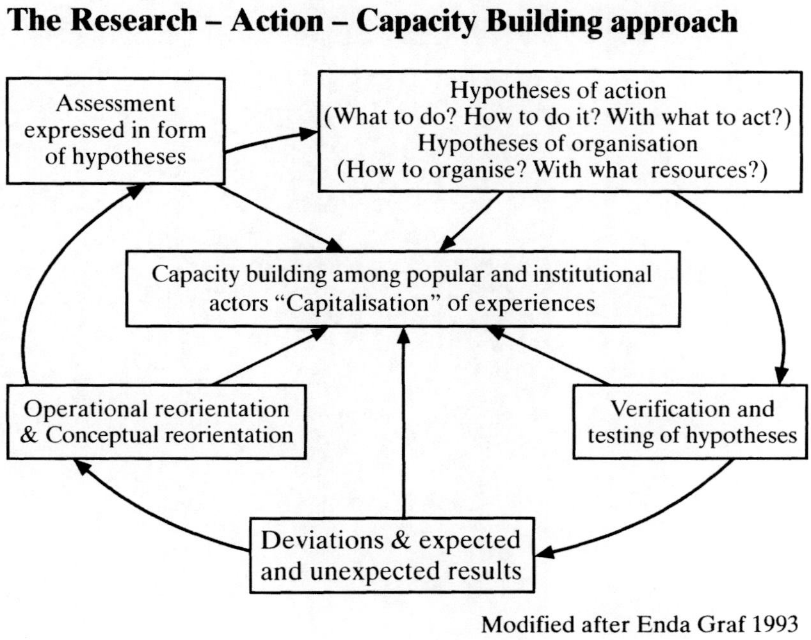

Figure 1
The Research-Action-Capacity Building Approach

Water and Solid Waste Management

In most cities of the South, access to water, especially drinking water, is a major concern. In order to describe the situation in N'Djaména, a household survey was carried out in three neighborhoods. This survey enabled researchers to quantify the problems (Amsler-Delafosse 1998). The results and additional research (Bachimon 1998) revealed that people spend a substantial part of their household budgets on drinking water. Moreover, poorer people have to spend a higher percentage on purchasing drinking water than those who are better off. After these research activities, lobbying through regular meetings and contacts and interaction between associations and donor agencies made it possible for associations to improve access to drinking water through installation of hydrants in several neighborhoods of N'Djaména.

In parallel, the project studied how existing community initiatives were operating in the field of waste management, and how these initia-

Figure 2
Community Initiative of Waste Management

tives could be made sustainable. The situation at the city level was assessed and results used in organizing a workshop that brought together main stakeholders at the city level. This workshop resulted in establishing a master plan for waste disposal for the whole of N'Djaména (Doublier and Dobingar 1998). As part of this plan, several associations were assisted in dealing with organizational, financial, and administrative issues related to composting degradable waste and improving waste collection procedures from the household level to intermediary depots. As a result of these efforts, solid waste management improved (Fig. 2).

Social Mobilization around Street Children

The growth of economic and social disparities, coupled with contradictions within urban societies, has produced new actors. Among the most visible are marginalized groups such as "street children." In N'Djaména and Dakar, social work in an open milieu was the approach used to prevent exclusion of children and young adults living on the streets. They were contacted and accompanied by members of the project team in their daily-living situations. To tackle health-related problems, the health status of street children and their access to health care was initially assessed through interviews, observations and group discussions (Levaque 1998). Feedback of results to health workers and resource persons of the Ministry of Health made it possible to inform and sensitize them to the importance of the issue. These exchanges also resulted in two dispensaries in the city agreeing to treat the children. A facilitator was appointed to deal with formalities and keep records for orienting and guiding children coming to the dispensary and follow-up.

Unfortunately, this solution did not prove sustainable, mainly because health workers were reluctant to treat street children. The children, in turn, mistrusted the providers. Subsequently, other solutions were discussed and implemented. A missionary-run facility was willing to welcome the most urgent cases temporarily. A physician and a health service also began to organize visits and treatment for the children directly in "their" places. Through this approach, and permanent availability of drugs, health problems could be tackled. This solution has proved sustainable. In this instance, project activities have catalyzed a process for identifying feasible operational solutions (Nodjiadjim and Wyss 1999). A communication and negotiation process, plus testing various solutions followed by reorienting activities, resulted in more definitive solutions.

Action-research activities have further strengthened contacts with donor agencies (UNICEF and Fastenopfer Switzerland). These contacts have enabled one partner association – APPERT – to establish community centers in various neighborhoods in N'Djaména. These centers provide schooling and job training possibilities, enabling steps toward reintegrating marginalized children and young adults into society.

Promotion of Insecticide-treated Nets for Malaria Prevention

Utilization of insecticide-treated mosquito nets (ITN's) can make a significant contribution to control of malaria. Partnerships with two community associations in Ouagadougou and three in N'Djaména have been established to promote distribution and use of ITN's. As a first step, assessment of the situation was made together with these community groups. Research on people's views about malaria and problems related to mosquitoes and nets showed that malaria is perceived as an important health problem. This fact was highlighted by people's investment of substantial resources to protect themselves (Nadjitolnan 1999).

Subsequently, centers for the sale and impregnation of mosquito nets were installed by the associations at the neighborhood level. These centers have encountered a number of organizational and economic problems. Solving them has required some readjustment of activities. It became clear that promotion of insecticide-treated nets has three aspects: technical, economic and social. On the technical level, an innovation could be introduced successfully and adopted by local actors organized into associations. On the economic level, sustainability of sales and impregnation centers proved to be fragile, mainly due to high prices of nets and impregnation services. The majority of the urban population could not afford them. On the social level, the RAC approach showed that community groups could play an active role in promoting ITN's, and that groups could generate their own ideas and come forward with sound proposals for creating centers. In other words, experience could be "capitalized."

Additionally, the project demonstrated benefits of collaboration within the project and between a government program (Malaria Control Program Chad) and non-governmental institutions, as well as community groups for strengthening relations and exchange of expertise in South-South and South-North collaboration.

Conclusion

The Research-Action-Capacity Building approach used by the project revealed the potential of this approach in three areas:

(1) development, identification, implementation and readjustment of community-based answers to urban environmental problems, for example, attempting to prevent exclusion of street children and young adults through social work in their natural milieu;
(2) collaboration through participation;
(3) acquisition and exchange of knowledge, and change and development at micro- and meso-levels, resulting in "capitalization" of experience.

The project also showed how collaboration, partnership, and communication between community members and institutional actors can promote sustainable management of the urban environment. Key events in the process of action research were regular meetings and workshops at local, regional and international levels with all partners and actors (residents, researchers, municipal planners and administrators, donors). These events provided an efficient platform for exchange and discussion of findings, discussion and, if necessary, readjustment of activities and joint assessment of the consequences. More importantly, they created an interface between donors and organizations and the local community, enabling the former to assist people living in the city in developing initiatives and finding and carrying out their own solutions. The results indicate this process can lead to interventions and activities that have a high chance of being sustained, despite prevailing constraints and the rapid pace of urbanization. Therefore, a transdisciplinary approach which brings together a wide variety of actors, can contribute to a more sustainable urban environmental management.

6

Outcomes

6.1 Reflections on the Interactive Sessions: From Scepticism to Good Practices

by Kirsten Hollaender, Research Institute for Sociology, University of Cologne, Germany and Pieter Leroy, University of Nijmegen, The Netherlands

Abstract
Five conference planners offer impressions of the meeting and lessons about transdisciplinarity's conceptual meaning, added value, and the involvement of non-scientists. Beyond the knowledge base of the conference, four areas need further work: theoretical and epistemological aspects, quality management, ongoing monitoring and self-evaluation, and political aspects. Promoting transdisciplinarity will require patience, involving the younger generation and a genuinely open, sustained dialogue with society.

Moderator of Panel
Julie Thompson Klein, Wayne State University, USA

Panelists
- Britt Marie Bertilsson, MISTRA, Stockholm, Sweden
- Alain Bill, ABB, Switzerland
- Kirsten Hollaender
- Pieter Leroy

Introduction

On the third and final day of the conference, a panel reflected on two very crowded days of presentations, speeches, and conversations. Our "morning-after" reactions were spontaneous, based on immediate impressions of the most striking moments, lessons, practices and issues. Reflecting on the vast array of contributions and discussions is a daunting, if not impossible, task. The twelve Dialogue Sessions (D), eighteen Mutual Learning Sessions (M), and fifty-five contributions to the Idea Market (I) resulted in a dynamic conference that captured the vitality and plurality of transdisciplinarity. As a result, our task was far more difficult than reporting on a traditional conference in a well-established field, where the outcome might be anticipated.

Beyond its subject – transdisciplinarity – the conference was distinct in another respect – the experiment of a new kind of forum. The Mutual Learning Sessions were not only co-organized with partner institutions from outside academia. They also attempted to initiate transdisciplinarity within the meeting itself, by stimulating co-production of socially-relevant knowledge among scientists, a variety of stakeholders and lay people. In addition, non-academics were active participants. In short, the conference itself had the character of an "in vivo" transdisciplinary experiment. In terms of participants, procedures, and outcomes, it was more gradated and open than most meetings.

Rather than attempting an impossible summary of every session, we have grouped our reflections around three major topics and related questions, with added insights captured from the audiotape of our actual panel discussion. The conference furthered the current debate on transdisciplinarity, what it is or can be, and how it should or should not develop. The three main topics are framed by that debate:

- Transdisciplinarity is a sensitizing concept, as responses to the call for contributions made clear. Yet, the concept provoked initial uncertainty and scepticism, raising questions about its nature, added value and relationship to similar concepts.
- Transdisciplinarity has been touted recently as a new concept. However, it is not entirely new. Co-production of knowledge relevant to tackling societal problems is already practised in a variety of participatory procedures and methods of knowledge production and implementation. These practices occur in both Southern and Northern countries, in politics, in business and in civil society. Consequently, transdisciplinarity should be viewed as a series of both existing and newly-developing practices related to the interface of science and society.
- Transdisciplinarity, as both a concept steering scientific efforts and a series of social practices, still needs to be furthered and elaborated. The conference left us with a tough set of challenges to undertake in the realms of epistemology, quality management, and organization and management of political processes.

"What was your favorite memory or image of the conference?"

Britt Marie Bertilsson:
"Monsieur Kleiber gave a speech on Monday morning in which he gave a new meaning to the expression '*ménage à trois*' – three strange bedfellows of science, market economy and democracy. While I was packing this morning, I tried to see their faces, but I couldn't see them."

Alain Bill:
"When I was asked to make the link between industry and the conference board, I had to explain to my bosses what 'transdisciplinarity' means. That was a big exercise, very confusing at the beginning. But, finally we could show that in industry we had been tackling that issue without using the word 'transdisciplinarity,' so to speak."

Kirsten Hollaender:
"I saw people being very enthusiastic yesterday after the Mutual Learning Sessions, including many professors. They cannot be surprised so easily. I also organized a dialogue session on the first day. We had so many papers, I had a radical solution. Everybody only got 5 minutes. And, it worked. People liked the idea, and we had 90 full minutes of lively discussion."

Pieter Leroy:
"Right from the beginning I found the word "transdisciplinary" very, very sensitizing, and obviously it is sensitizing, since it mobilized so many people from many, many disciplines from a variety of institutions."

Transdisciplinarity: What's in a Concept?

Like the participants, members of the conference planning board initially asked themselves what "transdisciplinarity" means. Our shared scepticism reflects some of the "popular objections" that *Jaeger et al.* identified in *D06: Risks and Uncertainties*. Scientists are also asking themselves what transdisciplinarity can add to existing concepts such as "interdisciplinarity" and "multidisciplinarity." Across sessions and plenaries, as well as the larger discourse of science, transdisciplinarity seems to be a promising new approach towards solving complex problems of societal relevance. However, basic questions arise. What does it stand for precisely? Does it really provide the promised extra problem-solving capacity? And, if so, under which circumstances?

Conceptual issues have been addressed elsewhere in this book. We agree on a central point of definition: the main characteristic of transdisciplinarity is analyzing and solving problems with substantial societal relevance. Hence, not only valid and reliable scientific knowledge are required. As *Gibbons and Notwotny* emphasized earlier (*Chapter 3.3*), transdisciplinary knowledge must be socially robust and acceptable, implying co-production of knowledge through cooperation among scientists, politicians, market economists and representatives of civil society.

Beyond common understandings, three different formulations emerged across deliberations:

- the nature of the concept and its relation to multi- and interdisciplinarity
- the involvement of non-scientists and ensuing implications
- the added value of transdisciplinarity.

Trans-, Multi- and Interdisciplinarity: A Conceptual Avalanche?

Basic definitions of the concepts of trans-, multi- and interdisciplinarity were compared earlier in several chapters. Here, we call attention to an implicit undertone of judgment, implying a hierarchical ordering of value. Many participants treated transdisciplinarity as better than interdisciplinarity, which was viewed, in turn, as superior to multidisciplinarity. Monodisciplinarity, in moments of exaltation, was cast as a contemporary pariah.

We join *Christian Pohl*, who contributed to *D01: Theory*, and *Jaeger et al.*, who contributed to *D06: Risks and Uncertainties*, in rejecting the idea of a hierarchy. Each approach has its purpose and merit. Moreover, neither transdisciplinarity nor monodisciplinary research is an end in itself. They are both scientific approaches for analyzing and solving problems. The most appropriate stance in any given situation depends on the character of the problem at hand, its history, its societal and scientific context, and the knowledge interests that are at stake.

Transdisciplinarity aims at a better understanding of complex societal problems by mobilizing and increasing society's reflexive capacities, thereby enhancing the design and implementation of solutions. The list of Mutual Learning Sessions – with topics ranging from genetic engineering and radioactive waste management to sustainable tourism and landscape development – underscores the urgent need for transdisciplinarity.

At first glance, and from a scientific view, the topics of papers might seem narrow: for instance, decision-making processes about high-speed trains. However, these issues are socially and politically complex. Transdisciplinarity has a double value in such contexts. On the one hand, it enables science and society to deal with complexity. On the other hand, it enables production of knowledge that is scientifically reliable, socially robust, and politically acceptable.

Distinctions among multi-, inter- and transdisciplinarity also reveal an implicit and rather static view of disciplines and their interrelations. In many sessions, we heard reports on the emergence of new areas of knowledge on the edges of disciplines, such as biochemistry and biotechnology. The interface of biochemistry, biotechnology and physics, which includes micromechanics and nanotechnology, is a particularly exciting example. These and other developments were clearly influenced by societal demands and, hence, are trans-scientific in origin.

The real argument, we concluded, is not about definitions and distinct practices of multi-, inter-, or transdisciplinarity. It is about achieving an adequate interrelation between science and society. Transdisciplinarity is unique in not referring primarily to the content, body, or bodies of knowledge that are involved. It is predominantly about processes of production, use, and implementation.

Transdisciplinarity Means Involving Non-Scientists

The idea that transdisciplinarity relates to the science-society interface and to co-production of knowledge was one of the starting points of the conference. The Dialogue and the Mutual Learning Sessions showed heterogeneity in the practical implications of this starting point. *D06: Risks and Uncertainties*, in particular, revealed lack of consensus on two related questions.

First, can a solitary research project, even when it critically assesses disciplinary scientific knowledge, be labelled "transdisciplinary"? In other words, can scientists by themselves be transdisciplinary, without involving non-scientists such as lay people, politicians and market representatives? *Britt Marie Bertilsson* recalled a pertinent example in *D09:*

Energy. In a discussion of socio-technological studies of various solar technology concepts, one scientist might examine how concepts were marketed. Yet, is this not simply an example of social studies in one discipline?

The second and related question is whether transdisciplinarity implies, of necessity, cooperation with practitioners? Put another way, does a simple orientation towards "real-world" problems suffice? In *D06: Risks and Uncertainties*, *Abbassi et al.* and *Jaeger et al.* contended that a scientific orientation towards practice is possible by making it the "object" of inquiry. However, the majority of papers and presentations paralleled the organizers' starting point – Transdisciplinarity refers to specific processes of knowledge production that imply involving non-scientific partners.

"Start a 'real' dialogue with society."

Too often, *Britt Marie Bertilsson* lamented, scientists and other knowledge producers stop their involvement with society at the level of making a popular version of four to five of their latest scientific papers published in prestigious journals. Research must not simply be explained to the layperson in a one-way communication. A genuine dialogue must be started. We are moving from an industrial to a knowledge society and can hardly grasp what that will mean ten to twenty years from now. Knowledge will be THE competitive edge. Citizens outside universities will also be knocking on our doors, asking "What are you doing for us?" Scientists need to have a reply ready.

Discussing the issue is one thing. Analyzing and assessing experiences is quite another. A series of Idea Market presentations documented a variety of participatory methods and procedures. *Vahtar* is a noteworthy example. Two quite different sessions came up with added evidence. *D08: Research Programs* showed striking differences in the extent and organization of stakeholder involvement in scientific research programs, ranging from ad-hoc advice to co-funding research and actively co-producing policy-relevant conclusions. *M17: Participation* depicted a similar variety. Participatory processes covered a number of issues, ranging from single issues to long-term processes. They also differed in the number of actors who were involved and in the envisaged and actual impacts.

In some respects, the variety of examples in *M17* hampered discussion. At the same time, this plurality indicates that involvement of non-scientists requires specific procedural and organizational prerequisites. Moreover, procedures, methods, and organizational forms vary, depending on the phase and the aim of non-scientists' involvement and the issue at hand. Different procedures are called for in mediating conflicting interests than in consulting the general public with the aim of designing possible scenarios. Clearly, a lot of work remains to be done on understanding the circumstances in which particular kinds of involvement, procedures, methods, and techniques are appropriate.

A related question arose in other sessions. What is the best timing of non-scientific involvement? There was a general consensus that transdisciplinarity should depart from the old-fashioned DAD scenario in which solutions are Decided upon within closed political or business arenas, Announced publicly, then Defended afterwards. Transdisciplinarity calls for another model of decision-making and another style of governance that refigures the roles of experts and stakeholders. In *D09: Energy*, *Warren W. Schenler* suggested that not only should stakeholders participate from the beginning, they must also be kept interested and active over the entire course of the project.

"Keep stakeholders interested and involved."

Involving stakeholders in defining a problem, *Britt Marie Bertilsson* acknowledged, can be tedious and take time. The challenge is to sustain their interest and involvement. They are valuable in setting the vision, so need to be kept active throughout the process of creating and assisting with new results, contributing empirical-tacit knowledge, and the resulting enhancement of systems analysis. Don't think they should just be accommodated at the end. Keep them involved, try to drag them from ordinary daily problems, and help them to look into the future. Only then does a common knowledge base emerge.

Yet another question arises. What is the right time to bring in non-scientists, lay people and practice partners? Some members of the conference argued they should be brought in as soon as alternative scenarios and probable options are clear. Others claimed they should participate right from the problem-formulation phase. Scholars have developed techniques for participation at this stage. Apart from normative arguments for early involvement of people outside academia, there is a pragmatic argument. They contribute local knowledge and traditional expertise, bringing innovative insights into both problem formulation and solution. On particular method, participatory knowledge syntheses, insures inclusion of their perspectives.

Similar questions arise about who should be involved. Transdisciplinarity should go beyond the model of counter-expertise used in the 1970s and 1980s, evident in environmental impact assessment, risk assessment and the like. They represented a "modernist" style of decision-making and bringing in expertise. Transdisciplinarity calls for a more open, pluralist, and, some participants said, a "postmodern" approach.

"Remember the lesson of history."

History, *Kirsten Hollaender* urged, provides a good lesson. Back in the 1970s, there was much discussion of interdisciplinary learning and research for society. The OECD had publications, and there were conferences. Somehow the movement diminished. There were successes, with new university departments and programs being founded. But, it was only partially successfully. If we want to prevent transdisciplinarity from becoming "history," we must remember this conference is a starting point, not a single event. We need networks, institutional structures, and exchanges of knowledge.

Openness and pluralism, though, are not given. They must be organized. When looking at the huge variety of actors and groups involved in and claiming a stake in different processes, the most influential ones tend to be well-organized, well-equipped political and economic actors. Unorganized citizens tend to be absent. The uneven access actors have to scientific knowledge is not easy to counterbalance. In their Idea Market presentation, *Van Veen et al.* addressed the unequal position of different partners in the case of soil contamination management projects.

Additionally, and no less sensitive, is the issue of whether transdisciplinarity, either de facto or deliberately, can be or should be an instrument for communication, implementation, and acceptance. *D05: Communication and Participation* featured papers on how to create better public understanding of scientific results. *M01: Technology Sharing*, *M18: Intercultural Learning*, and *M20: Theory and Implementation* all called attention to the idea of co-producing knowledge as a precondition for transdisciplinarity, whether in companies, universities, or global interrelations. Many contributions in the Idea Market expressed the same idea: cooperation with non-scientific stakeholders is aimed at gaining broader commitment and support for implementation.

The subject also came up in *M01: Technology Sharing*, where the idea of double-added value was highlighted. Transdisciplinarity provides innovative insights and solutions, makes them more commonly shared and thereby facilitates easier implementation.

The Added Value of Transdisciplinarity

Part of the initial scepticism about transdisciplinarity is related to the question of added value. Some participants argued that a transdisciplinary approach is mainly, if not exclusively, adequate for environmental problems. That was in fact one of the basic assumptions of the organizers, based on experiences of the SPPE. Many issues raised in the Call for Papers and many sessions also dealt with the environmental topics of climate, water, transportation, energy, soil pollution, solid waste, and radioactive waste. Obviously, environmental issues need a transdisciplinary approach and are an excellent training ground, since they are complex in both scientific and social terms.

Emphasis on environmental issues, however, does not mean trandisciplinarity's added value is limited to those issues. The added value for industry, Alain Bill notes, is to use knowledge better to improve performance. Efficient and effective knowledge sharing can create competitive advantages. *M03: Sustainable Banking*, *M04: Nutrition*, *M06: Health Costs and Benefits*, and *M13: Landscape Development* provided ample evidence that analysis and the design of solutions profits from a transdisciplinary approach in many fields. A wide variety of problems are characterized by scientific and social complexity. *M02: Gene Seeds* focused on biotechnological innovations emanating from the field of genetic engineering, with related implications for environmental, economic, and social impacts.

M02: Gene Seeds and *D06: Risks and Uncertainties* might lead to another misunderstanding. Does transdisciplinarity typically apply to what has been called the "risk society," referring to the high risks of modern technologies? This assumption not only sounds a bit ahistorical, but, both sessions on *Intercultural Learning*, *D07* and *M19*, especially the paper by *Mebratu* (*D07*), suggested that transdisciplinarity has a long history in Southern countries. In *D07 Jabbar et al.* (see *Chapter 5.3.1*) also reported on a project that evolved from disciplinary research towards integrated resource management and transdisciplinarity. This evolution is a good illustration of added value, indicating that past failures often result from neglect of local knowledge and ignorance of cultural differences.

Like *Mebratu*, *Jabbar* and the majority of other participants, we do not want to restrict transdisciplinarity to a certain empirical field. It is useful for a range of problems. Complexity continues to be the keyword, not so much in scientific terms where cross- or interdisciplinary research may be better indicated, but in social and political terms. The conference provided ample evidence that under such circumstances a transdisciplinary approach has added value. Involvement of non-scientists or, in general, opening and enlarging the arena for discussion, is a key element in promotion, though not a guarantee that problem formulations will be both scientifically-reliable and socially-robust.

A number of papers underlined this point, while adding empirical evidence. In *M19: Quality Criteria*, the advantage of transdisciplinarity for exact definition and localization of problems beforehand was highlighted, not simply for its potential to improve problem solution. Practice would have the role of helping scientists to identify truly relevant problems. Similarly, in *M14: Solid Waste*, major problems in implementation of solid-waste regulation were attributed to not involving civil servants who would be responsible for implementation in the design. In this particular case, inter-organizational differences can be seen as "intercultural differences," resulting in misunderstandings and unforeseen side-effects that could have been anticipated through dialogue. In the Idea Market, *Caetano et al.*, developed a related argument that more adequate problem formulation promotes better decision making and the possibility of more effective problem solving.

A few caveats emerge, however. Optimism should not blind us. Even when one accepts the principle of added value, transdisciplinarity does not in and of itself reduce complexity. It may merely reflect it. Some participants asserted that transdisciplinarity would even generate complexity. Clearly, there is no panacea, no one best method. This warning does not deny transdisciplinarity's added value, though. Its contribution is primarily its capacity to recognize a complexity that might not be taken into account in conventional research and societal discussions.

In this more realistic view, the science-society clash tends to bring about new perspectives, problem definitions, and options for solution. A traditional French saying comes to mind – "du choque des idées jaillit la lumière." In order to be fruitful, dialogue between science and society should not be a shock. It has to be organized with appropriate procedures and methods in mind.

Transdisciplinarity also has limits. It has the potential to unravel complexity; to uncover, identify, and structure various aspects of a problem; and to come up with some options never before imagined. Yet, ambitions and pretensions might be pitched too high, especially when transdisciplinarity is touted as a method for conflict resolution. In their different ways, *D06: Risks and Uncertainties* and *M01: Participation* addressed this issue. In both cases, it was clear that one cannot expect a transdisciplinary approach to bridge deeply-rooted societal divides, let alone change existing power structures. While reflecting a still honorable, albeit somewhat naive belief in the power of enlightenment, that would be asking too much from the sciences or from society.

Transdisciplinarity at Work: A Variety of Practices

In addition to conceptual challenges and uncertainties, the topic of promoting transdisciplinary practices was ubiquitous. The variety of examples that emerged was one of the hallmarks of the conference.

The Variety of Backgrounds and Contributions

As we all know, most scientific conferences gather people from either a single discipline or a restricted area of research. Despite common backgrounds or interests, these meetings reflect some heterogeneity. The Transdisciplinarity conference did not simply double that heterogeneity but squared it. Not only did people come from about fifty nations, they came from almost all disciplines and specialties: from philosophy of science, law, and ethics to fundamental physics, ecology, and mathematics in geography. On top of that, they came not only from academia but also represented business enterprises, governmental bodies, research-funding bodies and non-governmental organizations. Some are engaged in landscape development, others in radioactive waste, risk insurance and agricultural innovations in developing countries. Despite these differences, they shared a common interest in transdisciplinarity. The discussions that ensued did not lead to a Babel-like confusion. To reiterate, they gave the conference the character of a transdisciplinary experiment.

This experimental character was readily apparent in the Mutual Learning Sessions. The concept of "mutual learning" was fundamental to the conference, because transdisciplinarity leaves behind the "modernist" conception of science as a reservoir that provides state, market and society representatives with scientific evidence. The most appropriate metaphor for the new conception of the science-society interface is "speaking truth to power." Transdisciplinarity takes another stance, encouraging coproduction of knowledge that will enhance both the quality and acceptability of the outcome. Mutual learning is not only a metaphor for the science-society interface in transdisciplinary theory and practice. It is one of the cornerstones.

"Don't forget the 'human' dimension and adjust your 'frequency'."

Alain Bill suggested an image for promoting transdisciplinarity. "We must shift from here," placing his hand on his head, to "here," placing his hand on his stomach. Transdisciplinarity has to be IN the person. We must not think only about definition and theory but also the "human" approach. We must also remember that communication is essential. People may have a common sensitivity to the value of transdisciplinarity but are not on the same frequency sometimes. Their frequencies have to be adjusted by establishing communication from the very beginning.

The vision of Mutual Learning Sessions as "laboratories" of transdisciplinarity was evident in the fact that they were coorganized and cohosted with partners outside academic. Some were hosted by business partners, such as ABB (*M01*), Novartis (*M02*), and the Zürcher Kantonalbank (*M03*). Others were coorganized by municipalities or other governmental bodies (*M05*, *M13* and *M14*), by organizations engaged in education (*M07*, *M09* and *M19*), in health care (*M06* and *M10*), and NGO's (*M18*). This variety, to repeat, reflects multiple stakeholders who, in turn, represent a diversity of knowledge demands and interests. Apart from the variety of organizers, a majority of Mutual Learning Session also involved other parties, such as environmental activities in *MLS 15: Radioactive Waste* and artists in *MLS 19: Quality Criteria*.

Clearly, various fields of knowledge and expertise can and do contribute to the theory and practice of transdisciplinarity. *D01: Theory*, *D04: Complex Systems and Team Processes* and *M20: Theory and Implementation* demonstrated that one can draw on important theoretical insights from sociology and philosophy of science, epistemology, and systems analysis. In the Idea Market, *Jeffrey et al.* and *Gandolfi* illustrated these contributions. Insights were also drawn from psychology, management sciences, social sciences and linguistics. In many cases, they were supported by new technologies and opportunities in the realms of information, communication and dialogue.

Not surprisingly in such a labyrinth of specialities, fields and domains, some called for a new metalanguage. In the Idea Market, *Lunca, Canali and Jaeggi* advocated metalanguage. Many others endorsed the idea implicitly. This is not surprising, since it is a familiar call. More than a century ago Esperanto was promoted to avoid misunderstandings, in particular conflicts between powerful nation-states. Since these conflicts were caused partly by divergent languages, a common language was presumably a tool for furthering internationalism.

More recently in the 1970s, a similar plea was launched by advocates of new interdisciplinary language capable of overcoming quarrels between disciplines. Yet, moral appeal and theoretical elaboration did not end disputes. This failure does not mean we should not elaborate concepts and paradigms that might bring about more convergence. However, we cannot realistically expect that a single approach will become THE transdisciplinary theory, or that a single metalanguage will solve all communication problems.

A Variety of Good Practices

Participants not only varied in scientific and working backgrounds but in their familiarity with and engagement in transdisciplinarity. To cite just one example, a sociologist of science, well versed in theory of transdisciplinarity but with little practical experience, met a consultant practicing transdisciplinarity on a daily basis with little interest in theory. They, in turn, communicated with a small-business employer who only recently learned that what he has been doing for quite some time is now called "transdisciplinarity."

Mutual Learning Sessions were occasions for bringing these people together and bridging their differences. Reports on these sessions and the more traditional format of Dialogue Sessions, validated claims of success. Some sessions, such as *M17: Participation*, involved a lot of people from similar backgrounds, with well-established traditions of communication. Others were innovative in bringing together a variety of people who had never met before in such a setting. *M13: Landscape Development* featured participation by local people. *M04: Nutrition* and *M05: Sustainable Tourism* not only discussed transdisciplinarity, but, in the latter case, practiced it through simulation games. The Idea Market presentation by *Ulrich et al.* also indicated how gaming can be used.

Like transdisciplinarity as a whole, mutual learning practices cannot and will not always be successful. In *M15: Radioactive Waste*, a fruitful exchange of views occurred. Yet, the deadlock that often arises was also clear. Transdisciplinarity may be the most appropriate approach, but success is never guaranteed.

"Remember the young people."

Kirsten Hollaender reminded everyone that younger generations must be actively engaged as well. She is now finishing a dissertation and looking for a job. Like her counterparts, she faces pressures of disciplinary specialization while maintaining a commitment to transdisciplinarity. They ask where their "home" is and fear being locked out of the ivory tower of their disciplines if they become permanent travelers between boundaries and fields. "My wish," she proclaimed, "is that there will be institutional structures that enable us to do both and support us in both."

Britt Marie Bertilsson underscored the importance of remembering the young people, expressing concern about the overabundance of grey hairs at the tables of industry, agencies, and universities. People in their 40s and 50s are doing research that will influence the lives of our children and students, but we don't involve them. "Get the young people in there, because otherwise they will be chasing you."

What's Next? A Tough Program!

Taken together, the Dialogue Sessions, Mutual Learning Sessions, and Idea Market comprise a valuable picture of conceptual and practical aspects of transdisciplinarity. Providing a platform for exchange of visions and experiences was not just an official goal of the conference. It was a powerful outcome. In our view, the conference succeeded in establishing a state-of-the-art. Transdisciplinarity is not just a sensitizing concept but an inspiring concept that relates to a variety of practices, whether or not everyone is explicitly using the label. By enabling exchange of ideas and experiences across all sessions, the conference contributed positively to two of its major goals – developing and promoting transdisciplinarity.

Having said that, though, we must confront an unavoidable question – What is next? The conference leaves us with a large program of activities to be carried out in order to further transdisciplinarity, in both theory and practice. Four aspects will be crucial:

Theoretical and Epistemological Aspects

Theoretical implications and preconditions are just beginning to emerge. One important question, repeated in the debate on interdisciplinary research, is how to combine and integrate knowledge from different disciplines. Transdisciplinarity complicates that question, since knowledge originates from multiple subspheres of science, politics, market and civil society. To note just one example, the insights of filtering technology and toxicology on the probability of cancer induction in the neighborhood of a waste incinerator must be combined with the pre-scientific feelings and anxieties of people in that neighborhood. Providing them with technical information is not enough. Neither is a hearing in the local community building.

D01: Theory and *M20: Theory and Implementation* also accentuated the need for systems analysis, and other appropriate methods and techniques, such as participatory or group modelling and scenario building. By themselves, these approaches alone do not solve the problem of dealing with various insights and different types of knowledge, all which have to be certified as scientifically reliable and valid, on one hand hand, while socially robust, on the other hand. This is why transdisciplinarity squares the problems already faced in interdisciplinary projects.

A further point bears mentioning. As one member of *D09: Energy* mentioned with regret, a transdisciplinary research project takes time away from the current hallmark of good science, namely to publish in peer-reviewed journals. There need not necessarily be a conflict, *Britt Marie Bertilsson* urged. New journals, in print and electronic form, may be founded for discussing theoretical aspects. A first step was taken with the announcement of the SAGUF network for transdisciplinary research at <http://www.unibas.ch/mgu/sagufnet.html>.

Quality Management

The previous example of the waste incinerator illustrates the quality problem as well. How should trandisciplinary practices be evaluated? A large number of contributions explicitly addressed this challenge, in most cases conceiving it as an enlargement and complication of familiar procedures. Evaluation, of course, is an important issue for scientists, research management, funding institutions and, we would add, all co-producers and consumers of knowledge.

Familiar research evaluation criteria and procedures were widely regarded as a starting point. Yet, concepts, procedures, and criteria that go beyond traditional considerations in evaluation and peer review are needed as well. The challenge is twofold. It involves, first, assessment of the societal relevance of research and related efforts and, second, design and implementation of flexible evaluation procedures that acknowledge the diversity of transdisciplinary practices.

Both steps, moreover, should be established in a transdisciplinary way, as the result of joint development and agreement by the evaluators and the evaluated. *D03: Scientific Quality and Evaluation* contained good examples. *Spaapen et al.* accentuated different profiles of research related to different societal contexts, acknowledging the plurality of *practices*. *Defila et al.* introduced a new questionnaire tool for evaluating research proposals. On top of that, the SPPE has published reports on systems of evaluation at both program and project levels.

Even with the insights furnished by these presentations, designing a system for evaluating transdisciplinary initiatives remains an unfinished project. Furthering the idea of transdisciplinarity and, especially, getting it accepted by government, market and civil society representatives requires a more fully elaborated system of evaluation and certification. Without such a system, contemporary societal mistrust of science and technology will increase, not decrease.

Ongoing Monitoring and Self-Evaluation

Quality evaluation conventionally relates to more or less institutionalized and formalized procedures, aimed at assessing the quality of past performances and, when necessary, reallocating money and other resources. Yet, the conference also made clear that transdisciplinarity is among other things a process of social learning. Apart from formalized evaluation, we need ongoing monitoring, further exchange of experiences, and self-evaluation of experiments and initiatives.

Continued monitoring should include paying attention to several questions. Under what circumstances – scientifically, socially, and politically – is transdisciplinarity an appropriate approach, and when is it not?

When and how should non-scientists be involved? What are good practices of knowledge co-production? Continued monitoring and exchange of answers to these questions will aid in building a common body of knowledge and of tested practices.

"Build on the knowledge base established at this conference."

We must build on the common knowledge base established at the conference, *Pieter Leroy* exhorted. There are many levels where we could act, *Alain Bill* suggested. We must continue presenting examples of best practices and case studies. We must also keep the dialogue going. The number of participants at this conference, he proposed, means we should meet again, perhaps in two years, to see where we're at.

In *D04: Complex Systems and Team Processes*, *Loibl* emphasized the key role of team processes for social learning. *Schübel* highlighted the role of monitoring and team-coaching. These and other contributions affirm the need for a more or less permanent network and exchanges among practitioners of transdisciplinarity. In all countries, *Britt Marie Bertilsson* observed with Sweden particularly in mind, discussions about the necessity of academic outreach might benefit from a transdisciplinary approach. Discussion would imply more than just stakeholders' dialogue. When transdisciplinarity really implies is a major change in the science-society relationship. The conference was but a small, albeit important first step in forging a new relationship between science and society.

The Political Aspects of Transdisciplinarity

Despite its merits, the concept of transdisciplinarity runs the risk of becoming another buzzword, like "sustainability." The initial consensus about buzzwords, *Pieter Leroy* explained, typically reflects the vagueness of a concept. Later, the terms are used in a variety of circumstances to plea and defend, promote and mobilize. Inflation and devaluation, though, affect all managerial tools, including sensitizing concepts.

"Don't 'instrumentalize' transdisciplinarity."

We must not reduce transdisciplinarity to a mere marketing tool, *Pieter Leroy* warned. We must protect it from being instrumentalized so it can grow. This task can only be accomplished together, by doing and by reflecting on what you're doing so you can learn, so you can do things better in the future.

The danger of transdisciplinarity being reduced to a mere managerial tool was sounded in a number of papers, especially by *Grütter et al.* in *M10: Medical Knowledge*, *Kulikauskas* in *M11: Urban Quality*, and *Buchecker* in *M13: Landscape Development*. On one hand, managerial and mediating skills and tools are very important when it comes to bringing trandisciplinarity into practice. On the other hand, transdisciplinarity should not been reduced to a managerial instrument. Relatedly, a change is needed in the processes of knowledge production and consumption, opening them to former outsiders.

The need for change in knowledge production and consumption was discussed explicitly, almost by definition, in *M17: Participation*. Transdisciplinarity is inherently connected to democratization of access to and use of scientific knowledge. It represents an effort to redress uneven access to knowledge as a resource. This point takes us back to where we began and where we met each other.

The first Call for Papers sounded an important theme: "Knowledge is the only unlimited resource on earth." Making this a reality requires changing the processes through which knowledge interests are formulated and prioritized. They must be transparent and controllable. Ultimately, the political aspects of transdisciplinarity go well beyond individual case studies.

"Patience, patience, and patience."

What will it take most of all to success? "Patience, patience, and patience," were *Pieter Leroy*'s last words. "We have a lot of work to do!," *Britt Marie Bertilsson* added. But, she was reminded of the marathon runner, who has lots of patience and knows where he or she is running. It will take a long time to do what we must do, *Leroy* cautioned. Three things are of particular importance in the coming years: clarifying theoretical concepts and their value, building on the existing empirical evidence, and designing appropriate criteria, practices, guidelines, and directives.

Even with the "tough program" that lies ahead, *Alain Bill* responded, things are moving "in the right direction." Commenting on a remark from an audience member about lack of transdisciplinary education in schools and universities, Bill described his own company, ABB. It's clear that you must be both a specialist in your field and able to communicate across fields, breaking boundaries and walls. Increasingly, industry will recognize the "added" value of transdisciplinarity.

6.2 Learning about transdisciplinarity: Where are we? Where have we been? Where should we go?

by Roland W. Scholz, ETH, Zurich and David Marks, Center for Environmental Initiatives and Coordinator for Global Sustainability, MIT, Cambridge, Massachusetts, USA

Abstract

Scholz, a social scientist and decision theorist interested in theoretical basics, and Marks, a systems engineer with experience in real-world problem solving, offer insights on the epistemology, methodology, organization, societal and professional-practice dimensions of transdisciplinarity. Transdisciplinarity represents a move from science on/about society towards science for/with society. Their detailed recommendations include more work on theory, standards of quality, and creating strong institutional frameworks.

Introduction: Objectives of Transdisciplinarity

"Society has problems, whereas universities have departments." This common aphorism illuminates some of the current problems in knowledge production. The rising demand for supplementing contemporary modes of scientific production with "transdisciplinarity" is a signal that the efficiency of the science system, at least with respect to solving relevant problems of society, is insufficient. Transdisciplinarity complements the existing system with a new type of problem-driven, cross-disciplinary, cooperative approach.

As the aphorism suggests, progress with respect to societal efficiency is expected when scrutinizing the functionality of a discipline-based department structure. This is a delicate, highly ambiguous challenge, though, as Nowotny (1997) has indicated. Only a little more than 100 years ago, the aphorism had a completely opposite interpretation. At that time, departments were viewed as a means of progress in overcoming endless academic contemplation and as a potential way of providing scientific solutions to societal problems.

Three perspectives shape our view of the subject.

The Science Perspective

There are many definitions of "transdisciplinarity." In a 1973 OECD report on environmental education, it was defined as a state of knowledge production that occurs "when a common set of axioms prevail, related to but lying beyond and complementing traditional disciplines" (Emmelin). Occasionally, it means being very "interdisciplinary." However, a common understanding was evident in the conference. Transdisciplinarity is an activity that produces, integrates, and manages knowledge in technological, scientific, and social areas. As the prefix "trans" indicates, "transdisciplinarity" concerns go beyond disciplines.

There are three key components of definition:

a) supplementing traditional disciplinary- and problem-centered "interdisciplinary" scientific activities *by organizing processes to incorporate procedures, methodologies, knowledge and goals from science, industry, and politics*;
b) starting science production from *relevant, complex societal problems*, thus having the potential to contribute to sustainable development;
c) organizing processes of mutual learning *between science and society*, so that people from outside academia can participate in transdisciplinary processes.

(see *Scholz*; Häberli and Grossenbacher-Mansuy 1998; Jantsch 1972; Mittelstrass 1996; Giovannini and Revéret 1999; *Giovannini*; Kötter and Balsiger 1997, Nicolescu 1999, Wissenschaftlicher Beirat der Bundesregierung 1996).

Many of these characteristics are prevalent in applied interdisciplinary work. In contrast, transdisciplinarity goes systematically beyond common practices of applied and interdisciplinary scientific activity. It turns the procedure from the top down, beginning from the real world and screening the body of science for help. It also maintains a focus on problems. *Problem needs*, not scientific potentials, determine depths, degree, and extension of scientific activity.

The Societal Perspective

From a societal perspective, transdisciplinarity is a means for efficient utilization of bodies of knowledge in science, the technology domain, and other fields of society that control science by funding. Ideally, it seeks joint problem solving between science and society. Discussion of these issues raises a series of open questions that highlight the need to promote, organize, and evaluate processes and outcomes of current scientific practice. From the perspective of society, science is slow, careful, and performed for its immediate peers. Society's call for quick, relevant extensions often creates uneasiness about whether science will live up to its side of the bargain.

The Conference Perspective

The purpose of this conference was to look at process in organizing joint problem-solving between science and society. Transdisciplinarity does not ask for sporadic or temporary cooperation. Rather, knowledge integration should encompass joint problem definition, problem representation and problem solving. The concept of "mutual learning" is central to the process. This chapter addresses a number of pertinent issues, including how and why current structural and organizational changes are occurring; their necessities, potentials, benefits, risks and threats. We also review lessons about the three major goals of the conference: developing practices, promoting research, and creating favorable institutional structures and powerful incentives.

The chapter itself is a demonstration of mutual learning. The two of us are evolving to a similar viewpoint from different directions. Scholz takes the perspective of a social scientist and decision theorist interested in the theoretical basics of this new approach, with a focus on *knowledge integration*. Marks takes the perspective of an engineer used to real-world *problem solving*, in learning by doing and in pragmatic approaches. Together, these separate but converging approaches illuminate the roles of and need for knowledge production and application in problem solving.

Needs of Transdisciplinarity from a Social Science Point of View: Scholz

A Theoretical Foundation for Integrating Different Types of Data and Knowledge

According to the introductory definition, transdisciplinarity integrates different types of knowledge. A fundamental rule – *Without methods and methodology no science!*" – also holds true for transdisciplinarity. With that rule in mind, we must ask three questions:

(1) What types of knowledge are to be integrated in transdisciplinarity?
(2) Which methods are available for supporting knowledge integration?
(3) How can we accomplish the organization of such processes?

Expressed in other words, transdisciplinarity requires:

- Epistemology: a theory about what are the foundations, validity, and limits of data, statements and approaches to solving a certain problem;
- Methodology: a set of methods and a concept of how they are to be used, and
- Organization: management plans and processes of transdisciplinarity.

The necessity of epistemology was proposed in the conference by *Sundin* and *Burger*. *Burger*, in particular, raised the question, "How is it possible that one is able to formulate integrative knowledge on the basis of a ... diversity of different science grounds"? Expressing skepticism, he also argued:

> *Suppose it really is the case that there is absolutely no common base between different sciences in analogy that there is no common base in favor of a mutual translatable understanding between the communication of dolphins and our communication systems. Under this supposition it must remain totally mysterious how scientists from different "cultures" are able to work on integrative proposals.*

Burger's statement refers primarily to interdisciplinarity. Transdisciplinarity is even more tricky and complex, as subsequent examples in the field of regional or landscape development show.

Illustrations of the Necessity of Knowledge Integration

Let us consider two examples that illustrate the necessity of knowledge integration. Figure 1 documents the dramatic change of *landscape inventory* caused by loss of agriculture over the last several decades, due to increasing industrialization of agricultural production and urbanization.

1971
1978
1987
1994

Figure 2
Urban development and new technologies require an encompassing evaluation, which incorporates consideration of consequences with respect to different interests. As the ambient environment of residents and their contradictory aspirations towards life styles is affected, we have to consider not only quantitative or qualitative scientific data but also the different stakeholders' feelings and intuitions.

This sequence of illustrations can be considered from different perspectives, such as the number of species living in the environments, harvest yields, soil erosion or aesthetics. If problem solving is to be organized in this particular case, different *compartments* such as water, air, and soil system must be integrated, as well as *various disciplines* that embody knowledge about the compartments.

Figure 2 presents a kind of violent encounter of cultures. In this real-world planning project, the Golden Gate Bridge could span the Zurich lake. In a transdisciplinary process, the following question might arise: "By which means may we evaluate the appropriateness of this bridge?" Various disciplines can make the bridge safe and even aesthetically pleasing. But, connecting the two ends via the bridge will have major social, economic and environmental impacts. So, an evaluation process must deal with not only the cost and purpose of the bridge but also

---

Figure 1
An analysis of the rapid change of landscape inventory requires an encompassing knowledge integration with respect to compartments (water, soil, and air) and the incorporation of disciplinary knowledge from the natural and the social sciences.

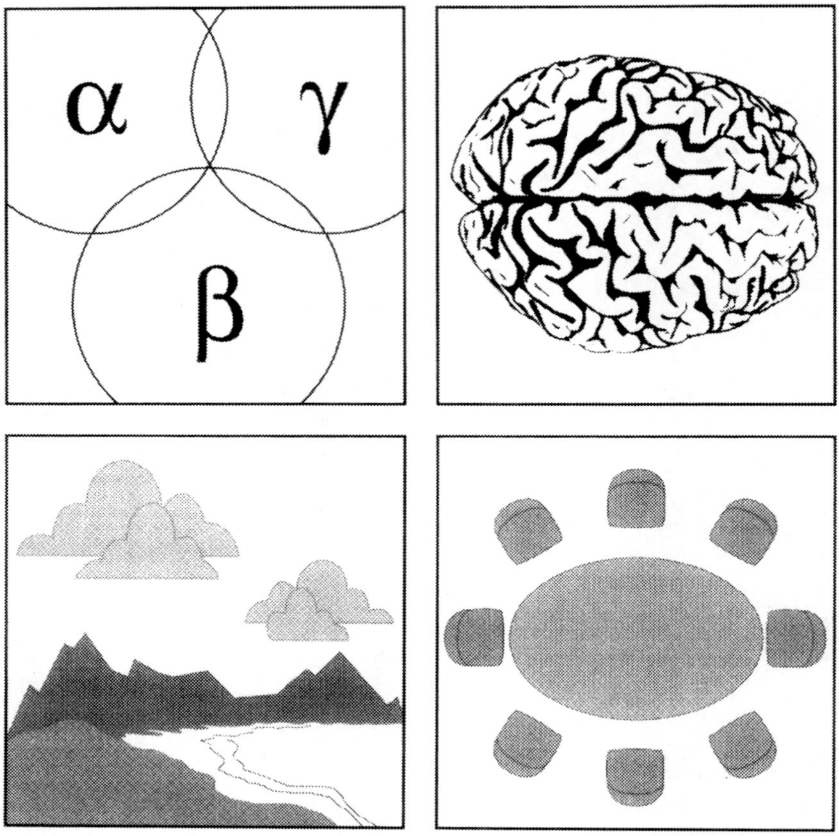

Figure 3
Transdisciplinarity requires methods that allow integration of knowledge with respect to at least four dimensions: (i) different disciplines in order to establish interdisciplinarity (upper left icon, α denotes arts, literature, languages, history, and philosophy/theology that are humanities and which were the initial subjects of university; β stands for natural sciences including medicine, and γ are the social sciences which entered the universities latest), (ii) different systems and compartments to allow an encompassing, holistic consideration (lower left icon, such as water, soil, air), (iii) different qualities of thought such as intuition and analysis (upper right icon, indicated by the right and the left brain hemisphere), and (iv) different interests of stakeholders (lower right icon, indicated by the round or oval table).

impacts, both positive and negative, that it will have on the linked communities and their regions.

As these examples suggest, transdisciplinarity needs methods for integrating at least four types of knowledge (Fig. 3):

The first dimension involves structure and procedures for systematically linking, or even fusing, knowledge from different sciences, for instance biology, economics, psychology and anthropology. These methods should establish interdisciplinarity.

The second dimension entails subdivisions into different (sub-)systems, which occurs in many cases. The landscape example may be considered from the perspectives of soil erosion, hydraulics, biodiversity or even climate change. However, compartmentalization does not help in real-world studies. In such studies, it is necessary to keep the whole, preventing over-compartmentalization and establishing a comprehensive, almost holistic view of a problem. Incorporating different views of participants in transdisciplinary discourse is very important. Each party has different values, preferences and issues in mind and will reconstruct the problem from a different perspective. Thus, one invariant of knowledge integration is mediation of interests.

The third and most important dimension is integrating different qualities of thought. If we consider the *Zurich Golden Gate*, as well as the other illustrations, common-sense knowledge provides an immediate evaluation. Different qualities of thought may be distinguished in the complementarity between *intuitive* and *analytic* modes (Scholz 1987). For mutual interchange between theory and practice, though, both types of thought are considered opposite. Intuition is not regarded as inferior or superior to analysis, but considered as a complementary approach.

In the case of the last type of knowledge integration, the focus shifts from methods or methodology to epistemology. Not only must integration occur between different language systems but the epistemic basis of the knowledge at work must be understood. In this context, the concept of an *architecture of knowledge* is helpful for ordering different types of arguments and reasoning inherent in transdisciplinary processes.

An Architecture of Knowledge

The architecture distinguishes between understanding (*verstehen*), conceptualizing (*begreifen*), and causal explaining (*erklären*) (Fig. 4).

To take the example of a landscape change (see Fig. 1), *verstehen* is on the top level, representing understanding of a real-world problem within its history, constraints, dynamics, and uniqueness. The key epistemics of cognition on this level are understanding by empathy, feeling, pictorial representation in memory and intuitive comprehension. At the second level is the conceptual model of the real world. A shift occurs from a

Architecture of Knowledge Integration

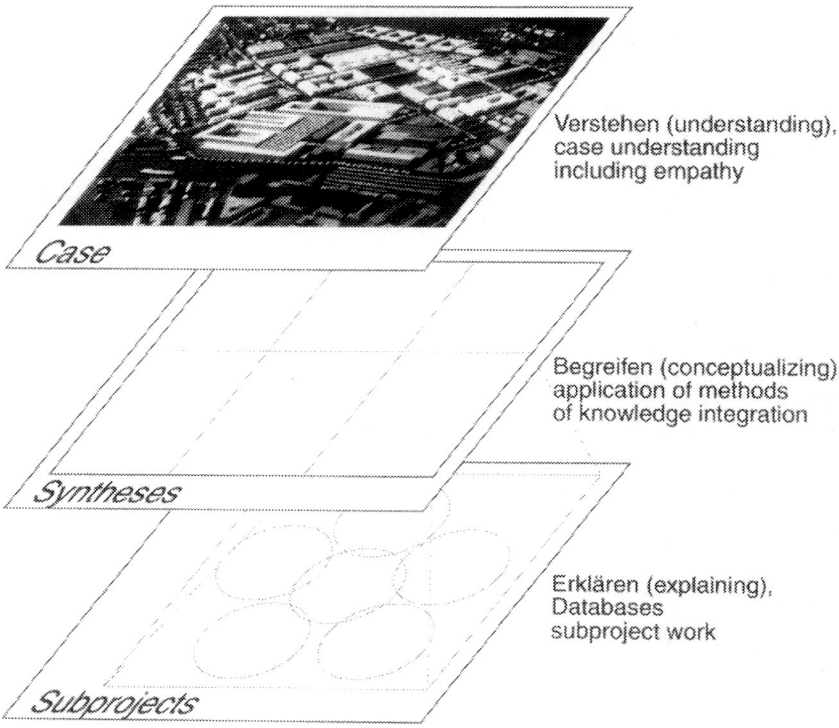

Figure 4
An architecture of knowledge. Three levels are distinguished: (a) understanding (*verstehen*) including empathy on a real-world level, (b) conceptualizing (*begreifen*) through synthesis by methods of knowledge integration, and causal explaining (*erklären*) based on arguments arranged according to propositional logic (Scholz and Tietje, in prep).

holistic real-world perspective, at the first level, to a system or model level, at the second stage. The key for successful work at this stage is synthesis and integration organized by methods of knowledge integration (Tab. 1). The third level is the epistemics of causal explanation by propositional logic, as prototypically provided by the formal natural sciences.

Thus, we have qualitatively different levels of knowledge that interact in a transdisciplinary process. To illustrate, one can distinguish at the top level intuitive understanding of a farmer about the dose-response

Table 1
Methods of knowledge integration. These methods allow for the integration of knowledge as presented in Figure 3. Most of the methods rely on a systems approach. The methods organize (i) problem representation, (ii) problem evaluation (iii) problem transition and (iv) study team organization.

	Method
Problem Representation	Formative Scenario Analysis System Dynamics Material Flux Analysis
Problem Evaluation	Multi-Attribute Utility Theory Integrated Risk Management Life Cycle Assessment Bio-Ecological Potential Evaluation
Problem Transition	Area-Development-Negotiations Future Workshops/Think Tanks
Study Team Organization	Experiential Case Encounter Synthesis-Moderation

relationship of a pesticide, on the basis of lifelong experience in farming, from highly analytic knowledge of an environmental toxicologist on the bottom level (Wynne 1996).

In the middle level, synthesis is an important step for establishing a scientific foundation of transdisciplinarity. Table 1 presents a preliminary list of methods considered as candidates for organizing knowledge integration for transdisciplinarity.

These methods emanate mostly from interdisciplinary fields such as decision sciences, operations research, and management sciences. Most of them have proved their value as tools for organizing synthesis in transdisciplinary case studies (Scholz and Tietje, in prep.). They can be classified as methods for (i) problem representation, (ii) problem evaluation, (iii) problem transition and (iv) study team organization.

This theoretical consideration ends with the example of the Theory of Probabilistic Functionalism, proposed by Egon Brunswik (1952) of the Vienna circle. This theory allows for a common conceptualizing of the methods and the epistemological considerations provided here. It refers to the sufficiency principle and vicarious functioning principle of information processing, providing a concept for generating robust conclusions and statements that are required from a societal point of view (see also *Gibbons and Nowotny, Chapter 3.3*).

Learning by Doing in Transdisciplinarity: Marks

I am an engineering systems person who has practiced all over the world on very large civil infrastructure problems with major societal implications, such as operation of the High Aswan Dam and water systems in the US, India, Greece, Argentina, and Colombia. In each case, the large-scale system must be analyzed from not only a technical view but also a social view. Knowledge from stakeholders about objectives, values, and uncertainties is as, or more, important than the detailed technical design of system components and their interactions.

This experience has given me an intuitive sense of the type of information needed, how it is collected and evaluated, how it is displayed and – most important – how to engage stakeholders in a constructive dialogue about choice and implementation. I also know Scholz's work and his theoretical approach to categorization. All this seems very familiar. Let us explore, then, the practitioner's view.

As a systems analyst, I am most comfortable with the paradigm of systems analysis problem solving. This paradigm entails:

- Problem identification – goals, constraints, objectives, measures of effectiveness, identification of stakeholders
- System model – being able to predict the results of actions taken in the problem solution
- Generation of alternatives to be considered for the problem
- Evaluation – using the systems model to consider the impact on system goals and on stakeholders of different alternatives and choosing a solution. (Notice the close link to Table 1)
- Implementation of the resulting choices.

Early in my career, engineering systems was largely a computer exercise, involving a great deal of attention to system modeling and generation of alternatives. I spent little time trying to really understand the problem (until it was almost too late). Moreover, the system model was often of a physical system behavior instead of resulting human behavior. Today, computation and models play only a small part. They have become standardized and fairly routine, especially with rapid advances in computational speed and efficiency. My current focus is understanding the problem better by building stakeholder processes that involve users of the final result in specification of the problem, its analysis, consensus building for solution choice and the lock-in for successful implementation. The emphasis remains practical, on the nature of the problem to be solved and on borrowing bits and pieces from new scientific, technological and policy developments. Improved practices in factoring in stakeholder involvement from the beginning are improving results.

Another question arises, though: "How different is this from the theoretical picture?," presented earlier. The answer is not very much. This approach simply starts at the practical end and evolves towards a common ground. Scholz presented a social science viewpoint that offers reflection and guidance for the practitioner of large-scale systems. The experience of practitioners offers insights into where new and applied developments are needed to advance the process. How interesting and revealing! The newly evolving science of transdisciplinarity already has a practitioner base that needs order and insights. It may be a new concept of "master engineering."

In former times the engineer was the master builder controlling the entire process. Now, in more complex systems, the engineer becomes the provider of analytic information for a stakeholder consensus-building process (Susskind, McKearnen, Thomas-Larmer 1999). Now, physical systems models have to add models of human behavior and societal dynamics and needs. Many solutions are not structural but require incentives for behavior modification. The practitioner badly needs guidance on how to predict this, an important insight as we see ourselves growing more closely together. Transdisciplinarity is a key formalization of this evolution.

What Have We Learned from the Conference?

The two of us sat through the entire conference and read all the materials. We offer the following recommendations from our different starting points.

From the View of the Practitioner: Marks

It was most interesting to observe a meeting of almost 800 mostly European scientists and people from practice struggling with definitions of things the practitioner has observed but could not classify. I was reminded of the traditional conflict between social scientists and practicing engineers. I spent many years trying to engage social scientists in my work but with little success. The practitioner is focused on the here-and-now of solving the problem at hand within time and budget constraints, rarely working on "academic exercises" even while acknowledging their value. The social scientist is trained to see across larger implications and to move from the specific to the general. The practitioner is concerned mainly with the minutia of the specific. This conference demonstrated that the two can move closer together under the umbrella of transdisciplinarity. However, a close community of transdisciplinarians entails some risks. Thus, I offer some warnings.

- *Beware of preaching to the converted.* This conference was composed mostly of people with similar perspectives, already well on their way to being believers in transdisciplinarity. In other words, they may be largely talking to themselves and are enthusiastic when everyone involved seems to "get it." These, however, are not the people to be convinced. We must move to the larger community that is much more skeptical, and to demonstrate that transdisciplinarity is a basic framework for better complex decision making and implementation. We must work on ways of widening this discussion and thereby adding to its vigor and expanse.
- *You must have a good systems model.* This is a basic precept of both engineering and my own underlying approach to knowledge integration. One cannot assume that something should be done simply because it seems like the right thing to do. For instance, arbitrarily limiting CO_2 emissions without regard to whether there are available technologies and/or proper incentives for changes in consumer attitudes in energy use and conservation is not a good idea, though it was an assumption in some of the discussions here. There must be a good systems model that shows the benefits and opportunity costs of too little or too much investment or of social restrictions that are not implementable. So, discipline yourself to think about how to show a particular solution *will* work, not *should* work.
- *Not all stakeholders value things the same way.* There was considerable discussion of the precautionary principle – do not do something before there is definitive proof that it points in the right direction. Some stakeholders in the sustainable development process, particularly among the richest, can choose to act in this manner. But, it may not be the best solution for poorer societies in the South. This difference in values between the haves and the have nots is at the basic root of the friction in moving towards global sustainability. All people do not have the same values, nor should they. Sensitivity to retaining local cultures and viewpoints should not be lost in the move towards globalized

markets for Northern products and, more important, Northern perceptions of the world. Be sensitive is the watchword.

- *Keep it simple.* To engage the practitioner you must communicate ideas in ways that will be understood. It is best at some point to look at a specific problem and see how social science is applied. And vice versa. To engage the social scientist, the practitioner must not dwell on past attempts but must look beyond to ways of shifting thinking in both camps toward a consensus on approach. This is best done by keeping things as simple as possible.

From the View of the Social Sciences: Scholz

The social science perspective reveals aspects inherent in the process of transdisciplinary activities.

- *Groups and networks have potential flaws.* Transdisciplinarity is a cooperative activity organized into groups. But, groups do not always perform better than individuals (Hare 1996). The same warning holds true for networks. *Von Reding* echoed this caveat at the conference, advising "We have to avoid the 'dark' side of networks: Network without hierarchy is austere; network without hierarchy is anarchy. Together they form a natural tension in the dance to discovery."
- *Science needs independence. Transdisciplinarity risks social contamination of scientific activity.* The independence of science was a major achievement of history. Evolved standards have stood the pressure of time, commercialism, and expedience. Will transdisciplinarity, with its desire to solve problems quickly, weaken this achievement? Similar exercises in the past have been dangerous in this regard. Science must keep its role and can participate in both camps. In the future, we are likely to see both transdisciplinary specialists, who work on the practice and science of this field, and other scientists, who move occasionally from their science role to participating in transdisciplinary process then back again.
- *Respect the "otherness" of other forms of science production.* Transdisciplinarians must take care not to pretend they are more important than disciplinarians. And vice versa. There is a tendency to do this in bringing the new into a well-ordered older disciplinary process. There is room for both, with good regard for one another. Transdisciplinarity is best conceived as a crossdisciplinary methodology to organize joint problem solving between science and society.

- *Do not create a discipline of transdisciplinarity.* As stressed here, development of methods for organizing transdisciplinarity is a crucial activity. However, transdisciplinarity should remain a subject in all disciplines.

Conclusions

Many outcomes have been already reported. In summarizing the state of the art, we end with the following conclusions. The general motivation for transdisciplinarity moves from science on/about society towards science for/with society. We also want to remind readers that the conference focused on a specific perspective on transdisciplinarity. Although there were many representatives from developing countries, the current Western technological view dominated. Variants of transdisciplinarity focusing on explication of humanity and human rights issues (D'Ambrosio 1997; *de Mello*) and Eastern variants focusing empathy towards nature (*Haribabu*) were discussed. However, they were not at the heart of the conference, though they are in line with this conclusion.

The conference, the pre- and post-conference proceedings and communications during and after the conference can be considered a "living document," evidence that transdisciplinarity represents a new type of scientific activity. This type is characterized by acknowledging the need for contextualization, seeking strategies for overcoming fragmentation (*Scheringer, Jaeger and Elsfeld*), dealing with complex problems, including stakeholders in problem definition and other phases, and elaborating new evaluation criteria for scientific activity that promote sustainable development. To conclude:

- *There is a good practice of transdisciplinarity.* Good examples already exist in fields such as regional development, urban quality, and North-South partnerships. Most of these examples are shaped by participation and incorporation of stakeholders' values, interests, and knowledge. Successful participatory processes are characterized by mutual acceptance of the diversity of interests between different actors. New, promising fields of transdisciplinarity are visible already, as evident in earlier summaries of the Dialogue and Mutual Learning Sessions.
- *Transdisciplinarity is an appropriate approach to coping with complex large systems.* Society is endangered by the fragility of large systems such as energy, traffic, financial markets, water, and nutrition. Problem definition, problem representation, and problem solution in this con-

text require knowledge that goes beyond the scientific world. In this sense, transdisciplinarity can be conceived of a new type of engineering work, particularly within the framework of sustainability (see <http://cei.mit.edu/>). A methodological foundation of transdisciplinarity is beneficial, if not essential.

- *Transdisciplinarity is a cross-disciplinary methodology* (as is interdisciplinarity). Transdisciplinarity is best conceived as a cross-disciplinary methodology that organizes mutual learning and joint problem solving between science and society. From a science perspective, it is one form of knowledge production and has to be distinguished from disciplinary, interdisciplinary and other forms.
- *Theory and standards of transdisciplinarity are under construction.* There is increasing awareness that the processes, dynamics and products of transdisciplinarity must, fundamentally, be designed differently from those of traditional criteria in scientific evaluation. Criteria and performance measures for quality control of transdisciplinary research are under construction.
- *An institutional framing of transdisciplinarity is needed.* Formal or informal institutionalization for promoting transdisciplinarity are needed.

a) *Transdisciplinarity Evaluation Boards for "calculable" reviews*:
Today, transdisciplinary enterprises are mostly evaluated within a disciplinary structure. As a result, submission of transdisciplinary projects sometimes resembles participation in a lottery and depends mostly on the openness and liberality of reviewers. In order to organize a reliable practice of evaluation and funding, we recommend creating small boards of three to five members from science, business, and administration on a national and/or international level. These boards should organize and conduct transdisciplinary reviews that conduct evaluations of transdisciplinary projects for funding.

b) *Transdisciplinarity Award – Conferences as a forum:*
The Swiss Transdisciplinarity Award is an important means of defining good practice and providing orientations, examples, models, ideals and standards. It should be awarded on a continuing basis.

c) *Transdisciplinarity Colleges for hosting (temporary) Transdisciplinarity Laboratories* (Fig. 5):
We strongly recommend initiating a "Forum of Transdisciplinarity," in the form of a Transdisciplinarity College. It would be a place where practitioners and scientists can find a stable base for devel-

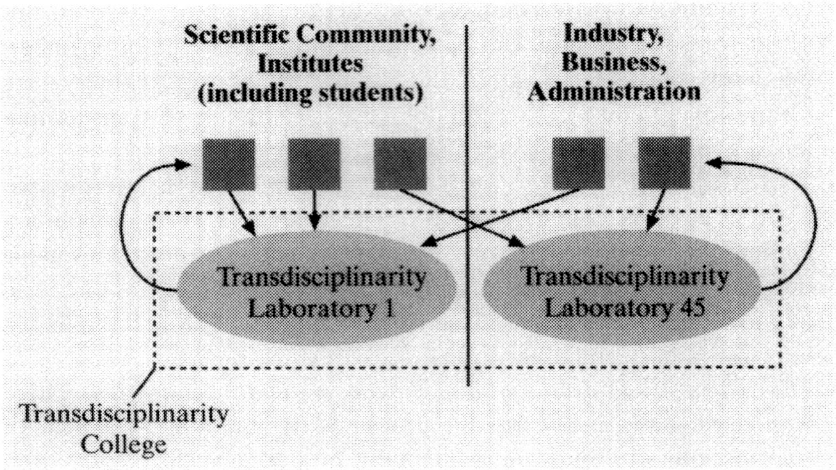

Figure 5
Design for a Transdisciplinarity College with temporary Transdisciplinarity Laboratories based on a symmetric participation of science and society.

oping, practicing and experiencing transdisciplinarity. This college should not have the structure of universities or permanent institutions. Rather, it should be comprised of Transdisciplinarity Laboratories, established for limited periods of time.

Depending on the problem to be dealt with, such labs could be formed for several weeks or even several years. In our experience it is crucial that the members show considerable commitment in time and motivation and that they return to their institutions afterwards. A symmetrical participation and liability of academia and practice for both program and finance seems indispensable. Clearly, the Transdisciplinarity College could also serve for young scientists as a platform for attaining reputations, and for qualifying for positions in academia and practice. This provision would minimize one of the greatest perils of transdisciplinarity: talented young scientists often avoid transdisciplinary projects because the fear of losing job career prospects. This impediment must be reversed.

6.3 Impacts on Science Management and Science Policy

by Hansjürg Mey, President of the Swiss Federal Commission for Universities of Applied Sciences, Berne; Sunita Kapila, International Development Research Center (IDRC), Canada-Branch Office in Nairobi, Kenya

Abstract
An international panel reflects on three areas: communication between science and society, education, and imbalance in the world between the generation of knowledge in the North and its application in the South, where easily-accessible forms of traditional technology and indigenous knowledge are valued. Transdisciplinarity can allow for integrating the two views and grappling with the complexity, interconnectiveness, turbulence, and rapidity of changes being experienced on local, regional, and global levels.

Moderator of Panel
Rolf Wespe, Head of Information Services, SAFEL Swiss Federal Agency for the Environment, Forest and Landscape, Berne

Panelists
- Hansjürg Mey
- Sunita Kapila
- Rainer Gerold, Director of Life Sciences and Quality of Life, Commission of the European Communities, Brussels
- Jan Eric Sundgren, President of Chalmers University Göteborg, Sweden

Key Statements
- Sell transdisciplinary products to those who formulate policy in order to convince politicians to base their decisions on scientific results.
- Learn in one day what can be said about transdisciplinarity in the media.

- Start better communication, new incentives, and transdisciplinary labs for young researchers immediately.
- Evaluate transdisciplinary projects in order to create new capacities and flexibility.
- Replace technology transfer in the third world by building up transdisciplinary capacities.
- Merge schools with different profiles to generate a self-acting process of transdisciplinarity.
- Transdisciplinarity is no substitute for monodisciplinarity.
- Young scientists should start in one discipline and, only after having reached a certain level, move into transdisciplinarity.
- The keyword for the 21st century is sustainability. Transdisciplinarity is one of the major tools to reach it.
- Transdisciplinary-like methods have existed a long time already, but naming, accentuation and diffusiveness are new.

How Can Science be Sold?

In traditional understanding, science was not supposed to sell. Science was just supposed to be. The new question "How can science be sold?" is primarily a question of communication. Communication between science and society is an old problem. It used to be a matter of financing, of getting the right research budget to convince the tax payer. We now have a new dimension: science should sell its products to those who formulate policy in order to convince politicians to base their decisions on scientific results. But, politicians depend on the public. They want to be elected. Therefore, scientists have to develop a strategy to influence politicians via the public. Many concrete examples illustrate this challenge such as the problem of food safety or genetic engineering.

What is the role of the media in this regard? Media has a lot of impact on the choices of young people today. But, you don't see engineers and transdisciplinary actors in TV. Television doesn't help the scene change. So, we have to make it more visible in the media and to overcome the basic conflict between communication and science. Research cognizance and technologies increase in time periods of decades. Politicians tend to think in time periods of four to five years, business typically in one year, and communication in one day. This mismatch of time constants has to be considered. "Science has to learn what it can say in one day."

Education

A lot of walls have to be broken down. We need to start better communication immediately with young people. This change must occur not only within universities but in high school and even earlier. What is the best way to change attitudes in universities? Get them actively involved and do not try to achieve something strictly top-down. If you enhance existing curricula with transdisciplinarity, you must start early, not just in the last semester.

One way of introducing "multi- and transdisciplinary thinking" is to reshape the education system. This change provides incentive to rethink traditional monodisciplinary curricula. New combinations lead to new academic and non-academic professions, such as medicine and computer science or management and civil engineering. One should seize the opportunity whenever change is provided.

One example is the tertiary educational system in Switzerland, with seven new "universities for applied science," which have regrouped over a hundred existing schools of varied types, such as architecture, technology, administration, arts, medical care and pedagogy. This merger of schools with very different profiles under the same roof, through integrative strategic planning, generates a self-acting process toward transdisciplinarity.

The Third World

We observe a severe imbalance in the world – between where science is generated and where it may be applied. The "first civilization" is based on growth of science as the main activity, creating a "Northern view." Science-related technology and its incorporation into production and social processes results in higher incomes. The "second civilization" is characterized by a low capacity for knowledge generation, though a broad base of traditional technology with a thin layer of imported knowledge on top of it. The result is low-income in 80% of the countries of the world. The "Southern view" finds value in easily accessible forms of traditional technology and indigenous knowledge that scientists have often swept aside indiscriminately.

Transdisciplinary perspective can allow for integrating the two views. At present, a significant proportion of science that is applied in the development context comes to the Third World through multinationals and donors. Our task as researchers is to develop transdisciplinary skills for adapting this science to local situations and stimulating development and application of indigenous science. This way of solving problems is now being valued by academic communities and those who previously tended to prefer monodisciplinary perspectives.

Problems are always multifaceted and multisectoral. What's new is that for the first time we are trying to articulate, to conceptualize, and to document ways of responding to problems in a holistic manner. This experience is occurring at all levels in many cultures. Discussions about transdisciplinarity and science and technology in the Third World are currently running on two levels: the North-South gap and the gap between scientific elite and the vast majority. Since the early 1970s, several donor-supported research programs such as the Canadian International Development Research Center (IDRC) have attempted to reduce these gaps by emphasizing participatory and transdisciplinary methods of knowledge and technology generation in Southern contexts. The vision behind their strategy is "Replace technology transfer by building up transdisciplinary capacities."

The need for interconnectedness in a fragmented world is everywhere. Southern researchers should work in Northern contexts in order to acquire expertise they can adapt to their home countries. The USA attracts brilliant scientists from countries such as China and India. However, Europe doesn't. Why not? We should improve our attractiveness to the most brilliant minds from developing countries, then encourage them to go back and set up institutions in their countries. More than communication is required. Responsibility must be shared by both the North and the South. In Africa "science policy" is simply words. It is not practiced in reality. Instead multinationals come in with their own Research and Development departments.

Evaluation Criteria

It should be easy to evaluate transdisciplinary projects: You have a problem, for instance the climate. You put money into this problem and somebody solves it. However, reality is much trickier and more sophisticated. Many stakeholders' perspectives and many practical issues are involved. Because transdisciplinarity is context-dependent, evaluation criteria are multifaceted and should fulfill the following requirements:

(1) Money should create capacity.
(2) Funds should be able to support requisite basic and applied research (the difference is not between basic and applied research, but between good and bad research).

(3) Consider the needs of key stakeholders.
(4) It has to have an output or be deliverable.
(5) Transdisciplinary work takes more time and more money as it covers a broader base of concerns. Give your people enough time (equally true for applied research).

The European Communities have vast experience in defining, evaluating, and funding research programs and projects. The evaluation procedure is divided into two steps: (1) evaluation of scientific merit and project management and (2) evaluation of social relevance. Both are executed by peers ("the other scientists") on a strictly anonymous basis. This prevents an unbalanced flow of money going to "famous researchers" and protects the possibility of brilliant newcomers receiving support.

The Swedish research administration applies the following criteria and measures successfully: (1) Support long term projects; (2) Look for clearly defined projects within a certain area; (3) Consider whether proposers establish a dynamic environment, such as transfer of research people in and out of different nations.

To promote transdisciplinarity, we must create new incentives for young researchers moving between boundaries. We have to establish flexible organizations with new stimuli, such as task forces and transdisciplinarity labs. We must adapt evaluation procedures and criteria to favor the potential of young researchers. We must allocate more funds to young researchers, perhaps investing in special risk-capital funds for them. And, we must give them an appropriate time frame, not just one or two years as is common now.

Evaluation-based quality control is a big issue today in all types of industrial and university activities, including education and research. Yet, it has an aura of fashion, buzzwords, and exaggerations. Researchers and professors complain about wasting their time with long reports. Exaggerated and misguided quality control may kill the creativity. Therefore, we must also establish a reasonable equilibrium between the need for administrative reports and the free zone of scientific and transdisciplinary work.

Conclusion

Monodisciplinarity versus *Transdisciplinarity*

The question of how close two or more disciplines should be to qualify as "transdisciplinary" remains unanswered. Another question also arises:

Is transdisciplinary research the modern or future type, replacing monodisciplinary research?

The answer to this second question is no. Transdisciplinary research is an additional type within the spectrum of research. Therefore, it coexists with traditional monodisciplinary research. More precisely, we need scientists who stay monodisciplinary and push the frontiers of science, with good results. Young students should start in one discipline then, after having reached a certain level, go into transdisciplinarity. If you are closed in one discipline and closed in one country, you should realize that the development of your scientific career is mostly beyond national boundaries and your first discipline. Hence we need much more mobility within Europe and beyond.

Criticisms and Comments from the Audience

- Unfortunately, no member of industry is on the panel. This is typical. The panel reflects the attitude of governments and educational institutions. In industry, money per se is not primarily the problem. Where we need transdisciplinarity is in spending the money. This is a major question: How do we spend money in a sensible and intelligent way?
- Transdisciplinarity should not be the keyword of the 21st century. It should be sustainability. Transdisciplinarity should be one of the major tools to reach sustainable life (a proclamation that received considerable applause).
- The feeling in the conference was one of sitting in a newly emerging church. It was not scientology, but "transdisciplinology" or something like that. This is a very dangerous situation: In essence, one uses new names for the same old god. This splits the world into two camps: those who believe and those who do not believe (another assertion greeted by applause). The goal is completely agreed on, but the naming must be criticized. The label suggests something is completely new, a thought expressed in many presentations at this conference. In truth it has been done for a long time.
- That said, the situation today is new: What we are grappling with is awareness of the complexity, the interconnectiveness, the turbulence, and the rapidity of change. In our individual lives, we didn't have this intense feeling before but people do now. Transdisciplinarity means going not only beyond disciplines but also beyond the walls of science and beyond the campus with its knowledge system.

- In the early seventies UNESCO started its program "Man and Biosphere," fostering research beyond the disciplines. The program's "biosphere's reserves" are working models of transdisciplinarity. Monodisciplinary research would by no means achieve the goals of the program. All who are involved in research should regard this biosphere program as much-needed primary sites of fieldwork. This belief should be considered an open invitation.

7

Conclusion

Mainstreaming Transdisciplinarity: A Research-Political Campaign

by Christian Smoliner, Federal Ministry of Science and Transport, Vienna, Austria; Rudolf Häberli, Swiss Priority Program Environment, Berne; Myrtha Welti, Swiss Foundation Science et Cité, Berne

Part 1: Is "Cracks in the Ivory Tower" Only a Metaphor?

Wide cracks are running along the massive walls of the ivory tower of science.

The conventional rules of how science works, transmitted over many years, are now being questioned by society. The roof of the tower has been removed, and the solid stonework of the walls is coming apart. The rooms of scientific theory, organization, and research practice, previously accessible only to experts, are now exposed to the eyes of a curious public. As a result, new space for action is opening up: space for novel forms of communication, cooperation and conflict settlement.

The scientific community is disconcerted. Are basic scientific values threatened? Will research be subordinated to a simplistic postulate of practical utility in the future? Is there a danger of a political instrumentalization of R&D (research and development)? Are these legitimate questions or merely unfounded fears?

The scientific community faces these challenges hesitantly but not without a strategic concept. "Transdisciplinarity" is a term that encompasses a variety of different forms of science whose objective is not only to gain knowledge but to stimulate innovations in the system of practice. New partnerships among science and politics, administration, the economic sector, and the public are being formed. Such partnerships forge a new societal agreement between "inside academia" and "outside academia." They are linked by a new thesis: we need less research for the public and more research with the public.

Part 2: Is Research without Practice Blind?

"Democracy requires transdisciplinarity," proclaimed *Charles Kleiber*, State Secretary for Science of Switzerland in his keynote address for the

International Transdisciplinarity Conference (*Chapter 3.1*). Kleiber confirmed the need of democratic societies for future-oriented guidance from science. Gibbons et al.'s (1994) notion of Mode 2 knowledge production embodies what is needed: "Transdisciplinary science reflects the real complexity of the systems"; "Transdisciplinary research results in socially robust knowledge." Transdisciplinarity connotes contextualization of science and the conceptualization of human beings by science. It also opens previously unused creative potentials by combining scientific knowledge with public knowledge rooted in experience.

These definitions confirm the usefulness of transdisciplinary research, a topic addressed throughout this book. They also underscore the metaphor of research and practice as a different sharpness of vision. The results of the Conference indicate that it will still take some time for the principle of transdisciplinarity to become accepted by the mainstream scientific community. Yet, representatives of Mode 2 research are already recognized as serious partners and competitors in the scientific system.

When the nearly 800 promoters, experts and "activists" of transdisciplinary research met in Zurich, these citizens of about fifty countries constituted a critical mass in terms of scientific creativity and practical experience. However, they are merely about 0.008% of the roughly ten million scientists active world-wide (Markl 1999). Not enough for an academic revolution? This is not the question. Power has never been a scientific argument. In view of this "David-Goliath" relationship, both modesty and a sense of realism are necessary virtues.

Three fundamental questions arise in this context:

- What has to be done to create awareness of the different rationales of science, politics, economy, media, etc. and to integrate them into the concept and project levels?
- What possibilities are there to support a fair negotiation of conflicting societal interests by scientific instruments and procedures?
- What is the role of participatory approaches, and how is quality to be defined in the continuum between independent and oriented research?

The question of how the different players of the innovation system, particularly science and politics, may promote evolutionary processes leading towards democratization of Research and Development (R&D) is of the greatest importance. The model of a "Mainstreaming Transdisciplinarity" campaign appears. The question of whether a research-political initiative with a socio-political focus is called for, or a socio-political initiative with a research-political focus, is at best of theoretical interest: Changes between both characters are preprogrammed.

Part 3: Who are Three Main Players of the Research-Political Campaign?

Scientists as Commuters across Frontiers of Research

There are no "transdisciplinarians" or "transdisciplinaritists" with mythical capabilities. Traditional scientists may dream of such a being. The actual propagation of transdisciplinary concepts, though, requires an alternative understanding of the qualities and skills that make it possible to do research of high quality by traditional standards within the changing contexts of practice. Technical, social and emotional intelligence must be equally important, in order to avoid being crushed between incompatible requirements of practice and science. Furthermore, transdisciplinary research does not take place in the protected environment of a laboratory, in which unsuccessful experiments may be repeated.

Successful transdisciplinary scientists are cross-frontier commuters in different worlds of knowledge and experience. They leave the ivory tower of science as a matter of course, have strong disciplinary competencies, reward interdisciplinary and intercultural action, have flexibility toward different contexts and partnerships, are characterized by sound management skills, attribute the same importance to communicating scientific facts in a non-scientific language as in specialized language, reflect critically on the political and social relevance of scientific work, and assume responsibility for implementation of scientific insights. Last, but not least, they highly value courage and the readiness to take risks.

Politicians as Partners and Opponents

Politicians do not take pleasure in pure science or the process of gaining scientific findings as such. Instead, they seek efficient solutions to concrete problems and a legitimization of political decisions and measures. These driving forces of political interest in research are frequently underestimated, especially by representatives of basic sciences. As a result, misunderstandings occur with momentous negative consequences and impediments to building fair partnerships. Mutual reproaches such as "Politics are impatient!" or "Science is inefficient!" are detrimental to realizing a sustainable cooperation between science and politics.

At bottom, politicians are partners who may be easily won over and whose behavior is calculable, provided that the rules of politics are observed and the reality of hidden agendas and interests is not ignored. Nevertheless, cooperation models conforming to ideal types may have a Janus face, resulting in negative long-term effects that often go unrecognized. Partnerships without the corresponding opposition are simply inconceivable in the political system. (If you decide in the political system in favor of one party, you are at the same time against another party – an opponent.)

Research and Development Managers Acting between Visions and Reality

Like the two above-mentioned groups of players, administration and public R&D management has its own institutional rationale. This rationale is characterized by the need to balance continuously the visions of subject-oriented politics, the reality of daily politics, and the framework of technical and legal requirements. It is an unpleasant sandwiching, wedging managers between contradictory expectations of politics and science. They must also respond to the self-confident efforts of transdisciplinary researchers to emancipate themselves from excessive bureaucratization of governmental administration. Extremely intricate and complicated guidelines for awarding research contracts and unrealistic evaluation activities that are abused to pave the way for budget cuts are only two examples of the obstacles.

The following analysis aims at demonstrating the profound changes administration must undergo if the objective of establishing transdisciplinary research in mainstream science is to be met. R&D administration praises interdisciplinarity but still promotes disciplinarity. A superficial glance at national research budgets reveals that disciplinary excellence is still the credo of funding. R&D administration demands transdisciplinarity, anxiously reducing the independent actions of science at the same time.

To cite a particular example, administrative regulations and methods of resource allocation still impair opportunities for young scientists to compete successfully with established R&D institutions, particularly university departments. Likewise, insufficient incentive systems for interinstitutional and international mobility have a counter-productive effect on fruitful border crossings in research and education.

R&D administration needs system-oriented knowledge and sectoral knowledge. This fact is not surprising, since administration is split. Analogous to the disciplinary division of science, administration is segmented into thematically separate units of ministries, departments and specialized units. In most cases, sectoral problem solutions represent a shift of problems to other areas of responsibility. They are hardly sustainable but promise short-term success.

R&D administration also calls for courageous scientists but fears, itself, powerful partners: transdisciplinary research may serve as an efficient control mechanism of administrative action, revealing sins of the past and questioning cherished but obsolete rules. Once again, the hybrid nature of transdisciplinary research is manifested. The welcome innovator of today may be a dreaded judge tomorrow.

Part 4: How Can the Campaign for a Research-Oriented Intervention be Realized?

Obviously, the internalization of transdisciplinary research by all those affected is a prerequisite for institutionalization. The model of a research-political campaign is well suited for this new form of science oriented towards action and intervention. The slogans "Through action to awareness" and "Help to self-help" can be guiding principles of intervention in science, politics and the public. The demands and objectives detailed below are derived from discussions at the conference. Their underlying aim is to provide the necessary motivation to initiate the campaign in one's own sphere of responsibility.

"From the Ivory Tower to the Innovation System"

- Research policies have to be integrated as a cross-sectional subject into all political spheres.
- Transdisciplinary concepts must not remain limited exclusively to applied research; transdisciplinarity is also required in basic research
- Partners of practice have to be involved and granted equal rights in defining and realizing oriented research in platforms of cooperation. Research must address the process of practical application of findings as a subject in its own right.

- Continuity, a factor highly significant for transdisciplinary research, has to be ensured by long-term research initiatives (up to 10 years and exemplified by the Swiss Priority Program Environment (SPPE; <http://www.snf.ch/SPP_Umwelt/Overview.html>) and the Austrian Landscape Research Program (<http://www.klf.at>).
- Transdisciplinary phases in professional careers have to have positive effects on careers in the scientific system.
- Extra-university R&D institutes must not be discriminated against vis-à-vis universities.
- We need a culture of transdisciplinary cooperation more than transdisciplinary professorships or institutes. An increased output-orientation and functional orientation of research policies is the basis for structural reforms in the domain of science.
- The creativity of the entire innovation system has to be made use of. Uncommon partnerships, for example, between scientists and artists or between researchers and pupils, may be highly productive.
- There is a pressing need to abolish bureaucratic obstacles at functional and institutional levels. High-risk projects require special support.
- Concepts combining transdisciplinary research with transdisciplinary instruction have to be developed; this principle also extends to educational offerings outside of universities.
- Public attention and appreciation of transdisciplinary research has to be promoted by targeted public relations campaigns.

"Transdisciplinarity Needs Internationality"

- Research in international project teams is increasingly regarded as the "better form of research." Theme- and problem-related networks of researchers and practitioners have to be propagated at the international level. In this context, modern information and communication technologies are indispensable. Innovative forms of cooperation in the form of global partnerships must be initiated, including virtual institutes or virtual centers of excellence in transdisciplinary research.
- International mobility programs for scientists have to be adapted to the specific requirements of transdisciplinary projects, including the time periods of projects and monitoring of results.
- Requests formulated by international institutions such as UNESCO, on the occasion of the World Conference of Science 1999 in Budapest, to strengthen transdisciplinary research must be met. International

conventions, such as the Convention on Biological Diversity, must also be utilized for transboundary research cooperation.
- The transdisciplinary component of EU (European Union) research policies and policies of other multinationals has to be strengthened by coordinated initiatives of member states. Targeted lobbying for transdisciplinarity by institutions of science and practice will foster both the basic will to change and interesting pilot actions of individual national administrations.

"From Global Uniformity to Diverse Transdisciplinary Cultures"

- This principle is closely connected with the above-mentioned one. Contrary to Mode 1 science that is more or less standardized internationally, the biggest innovation potential for transdisciplinary research is to be found in the globally enormous variety of cultural concepts of using expert knowledge and the public's knowledge based on experience. In the framework of concrete cooperation projects, above all between industrialized nations and developing countries, mutual learning has to be made possible.

"Least-Cost Science is an Important Concept in Transdisciplinarity" and "Transdisciplinary Science Needs Transdisciplinary Funding"

- Least-cost science is the dictate of the hour, not only in view of scarce funding budgets. The principle has been derived from the concept of least-cost energy planning of the energy industry. Applied to publicly commissioned research, it means that the aim is not a maximization of knowledge but the most effective use of research to optimize problem solutions (Loibl and Lechner 1997; Smoliner et al. 1998). The research-sensitive "acupuncture points" of a problem have to be treated. This is only possible if the following three types of knowledge are given equal consideration in transdisciplinary projects: system knowledge, target knowledge and transformation knowledge.
- Transdisciplinary funding means cooperative funding and requires a joining of funds with sectoral orientation along innovation processes or problem solutions. In the EU, an intensified cooperation of EU-funded research and activities financed through regional and structural funds has to be promoted.

- The requirements of practice often lead to unusual and expensive projects involving intensified communication needs among project participants. Funding providers must show increased readiness to accept and admit also these kinds of projects.
- Research projects require sufficient resources to make transdisciplinary learning possible. The project work of the practice partners also has to be financed. More seed money, planning money and risk insurance need to be provided than in the past in order to stimulate transdisciplinary cooperation.

"Transdisciplinary Science Has to Be Written into Society"

- Innovative findings are not yet innovations. Voluminous research reports are scientifically unattractive and not fit for practical applications. They should be replaced by target-group-specific products whenever possible. Besides the expert article that quite rightly is the central communication basis within the scientific system, there are varied alternative products for the work of researchers in the innovation system. The scope ranges from educational software, games and TV programs to works of art and technical problem solutions.
- Science marketing as well as intensive integration of users into the research process are the basis of research with social relevance.
- Clear agreements regarding intellectual property rights should be a prerequisite for every form of partnership.
- Long-term programs, such as the Swiss Priority Program Environment and the Austrian Landscape Research Program, are special forms of "learning organizations" that have a bigger impact on society than uncoordinated individual projects.

"Balancing Cooperation and Competition; New Ways of Quality Management"

- Transdisciplinarity is inconceivable without stable partnerships inside and outside academia. High scientific quality may not be achieved without competition. Therefore, innovative concepts have to be developed to link the antagonistic elements of cooperation and competition.
- Transdisciplinary science requires an initiative in theory and methodology.

- Successful research without tight project management is inconceivable.
- Preparing quality criteria for transdisciplinary research must become a top priority.
- Very often the societal impact of transdisciplinary research only becomes visible years after completion of the project work. Therefore, it is necessary to plan and to implement efficient monitoring concepts in order to determine the impact.

Appendices

Appendix A

Conference Organizers

Swiss Priority Program Environment

Swiss National Science Foundation, Berne

The Swiss Priority Program "Environmental Technology and Environmental Research" (SPPE) was started 1992 after the Earth Summit in Rio de Janeiro. The Swiss parliament commissioned the Swiss National Science Foundation to launch this research initiative with the following goals:

(1) to strengthen environmental research and to create priority centers and networks
(2) to stimulate cooperation in environmental research among universities, industry, administrative agencies and organizations
(3) to reinforce international research cooperation (EC and North-South partnerships) and to promote practice-oriented implementation of results in society and the economy as well as in politics.

The first phase (1992–95) had 120 projects and about 300 researchers, and the second phase (1996–99) 80 projects with about 200 researchers. The third and final phase of synthesis and implementation (2000–01) has 28 projects with about 50 researchers. Program managers oversaw organization of the entire undertaking and stimulated cooperation in environmental research. The program was also the initiator of the International Transdisciplinarity Conference. The transdisciplinary approach was a guiding principle throughout the life of the SPPE, which aimed to present a variety of examples and to promote experiences in using the approach.

Further information: <http://www.snf.ch/SPP_Umwelt/Overview.html>

ETH

UNS Professur für Umweltnatur- und Umweltsozialwissenschaften

The ETH/UNS

The Chair of Environmental Sciences: Natural and Social Science Interface (short ETH-UNS) was founded in 1993. It is one of eighteen professorships of the Department of Environmental Sciences at the Swiss Federal Institute of Technology. The name and affiliation of the Chair indicate the necessity of integrating social science into the environmental system. ETH-UNS focuses on analytical methods, case study methodology and implementation tools for transdisciplinary environmental decision making. Soon after its foundation, it became evident in the ETH-UNS case studies that the interface between science and society is at least of equal importance to the interface management between disciplines.

The ETH-UNS case studies cope with real, complex, societally relevant problems that are shaped by environmental problems. Examples of these studies are Sustainable Agriculture in the Grosses Moos Area (1994), Responsible Soil Use in the Klettgau Area (1998), Sustainable Urban Development of Zurich North Center (1996) and of Eco-efficiency of the Swiss Railway Company (1999). The ETH-UNS case studies are a hybrid, combining teaching and research in the field of transdisciplinarity. In each of the eight studies run since 1994, 60 to 120 students, about 25 scientists and 50 to 250 representatives from the case interact intensely for about three months in problem definition, problem representation, problem transition, etc.

ETH-UNS also has an active research group of fifteen scientists including about ten Ph.D. students from different disciplines running projects in various fields such as Sustainable Credit Management, Soil Remediation and Waste Management, Environmental Evaluation, Life Cycle Assessment, etc. The ETH-UNS research group was actively involved in the SPPE program and initiated a series of Mutual Learning Sessions for the International Transdisciplinarity Conference.

Further information: <http://www.uns.umnw.ethz.ch>

ABB

ABB: "Ingenuity at Work"

The ABB Group serves customers in power transmission and distribution; automation, oil, gas, and petrochemicals; technology building; and financial services. With novel Information Technology applications, tailored software solutions, growing eBusiness and a rapidly-expanding knowledge and service base, ABB is building links to the new economy. The ABB Group employs about 165,000 people in more than 100 countries.

The pace of technological development is increasing day by day. As new products and technologies are introduced, with profound impacts on many aspects of our lives, the growing complexity of innovation challenges us to question the efficiency and usefulness of each new development. In the past, we were accustomed to viewing technological development as an aid to improving standards of living. Today, we are increasingly alert to its impact on sustainability, efficient use of energy and environmental protection.

ABB – as a leader in several key industrial areas – is at the very center of these challenges. With its leadership role comes the responsibility to secure real and practical innovations. Our aim: to sustain growth and profitability through meaningful innovation.

The accelerating pace of globalization, combined with rapid swings in world economic markets, is redefining the previously established role of R&D in the corporate world. Economic changes, coupled with a significant increase in cross-border competition and enforcement of environmental regulations, requires complete dedication to technology-based innovation and to promotion of transdisciplinarity research.

Swiss Foundation Science et Cité

The Science et Cité Foundation was founded in Switzerland in 1998 by the four Swiss Academies (Social Sciences, Natural Sciences, Medical Sciences and Technical Sciences), by the Swiss National Science Foundation, the Silva Casa Foundation and the Swiss Federation of Commerce and Industry. Since 2000 it has partly been financed by the Swiss Confederation.

The Council of the Foundation is presided over by the Swiss Secretary of State for Science and Research. The purpose of the foundation is to re-establish dialogue between the world of science and research on one hand and society in general (i.e. culture, politics, economy, education, etc.) on the other hand. Not only should researchers and scientists become aware of their responsibilities both to communicate with the world "outside" and to respect people's concerns and expectations. Citizens (i.e. the people) should become aware of their right to ask questions and get answers and, for their part, should assume responsibility for scientific development. This dialogue should be the basis of a state of "critical confidence" between science and society. In addition to fostering better communication between the academic world and the world "extra.muros," the Foundation also tries to contribute to a better understanding within academic disciplines.

In its initial phase the Foundation organizes different platforms for this dialogue, i.e. "Café Scientifique" (discussions in public places on scientific issues), "Tables Rondes" (specific research projects regularly presented to a representative group of citizens), "Festival Science et Cité" (a week of Science presentations, discussions, expositions, shows and manifestations planned for 2001 in each University town and in Ticino) and a network of all those institutions (museums, etc.) which already work in the field of communication of science. In the longer term, the Foundation plans to establish Science Centers in each linguistic region in Switzerland.

Further information: <http://www.science-et-cite.ch>

econcept AG

is a private consulting firm in the field of environmental management and economics with a specific focus on sustainable development. econcept AG offers a multidisciplinary team of engineers, scientists and economists, plus comprehensive experience and expertise in transdisciplinary projects. econcept AG is collaborating in an extended network of private and public research institutions and coaches public and private administrations towards new forms of public management.

Marketing Service Ltd

is a full service agency for conferences and events on a Swiss as well as an international level, from the idea to the final event and its introduction into the market. Marketing Service Ltd, Zollikerstrasse 234, 8008 Zurich, Switzerland. E-mail: market@spectraweb.ch

Honorary Board, Organizing Committee, and Conference Board

Honorary Board

- Ruth Dreifuss, Swiss Federal Minister of the Interior, responsible for Research, Higher Education and Culture, Berne
- Alex Krauer, Chairman of the Board of Directors UBS and member of the Swiss Council for Sustainable Development, Zurich
- Augustin Macheret, State Councillor and President of the Conference of Swiss Universities, Fribourg
- Eckard Minx, Daimler Chrysler Ltd, Berlin, Germany
- Christian Smoliner, Federal Ministry of Science and Transport, Vienna, Austria

Organizing Committee

- Dr. Rudolf Häberli, Conference Chairman, Director of the Swiss Priority Program Environment, Swiss National Science Foundation, Berne
- Prof. Roland W. Scholz, Natural and Social Science Interface (ETH-UNS), Swiss Federal Institute of Technology, Zurich
- Dr. Alain Bill, ABB Corporate Research Ltd, Energy and Global Change Department, Baden-Dättwil
- Prof. Bernard Giovannini, Département de Physique de la Matière Condensée, University of Geneva
- Walter Grossenbacher-Mansuy, Swiss Priority Program Environment, Swiss National Science Foundation, Berne
- Prof. Ruth Kaufmann-Hayoz, Interdisciplinary Center for General Ecology, University of Berne
- Prof. Albert Waldvogel, Vice President Research, Swiss Federal Institute of Technology, Zurich

- Dr. Barbara Haering, econcept, Zurich
- Pius Müller, Marketing Service, Zurich

Conference Board

- Dr. Rudolf Häberli, Conference Chairman, Director of the Swiss Priority Program Environment, Swiss National Science Foundation, Berne
- Prof. Roland W. Scholz, Natural and Social Science Interface (ETH-UNS), Swiss Federal Institute of Technology, Zurich
- Prof. Maurice Campagna, Chief Technology Officer, ABB ALSTOM Power Ltd, Brussels, Belgium

- Dr. Philipp W. Balsiger, Interdisciplinary Institute for Philosophy and the History of Science, University of Erlangen-Nuremberg, Germany; Prof. Hans Beck, Président de la Commission de Planification Universitaire Suisse, Neuchâtel; Britt Marie Bertilsson, The Swedish Foundation for Strategic Environmental Research MISTRA, Stockholm, Sweden; Dr. Alain Bill, ABB Corporate Research Ltd, Energy and Global Change Department, Baden-Dättwil; Dr. Roman Boutellier, President and Chief Executive Officer, SIG Swiss Industrial Company Holding Ltd., Neuhausen; Prof. Max M. Burger, Friedrich Miescher Institute, Basel; Daryl E. Chubin, United States National Science Foundation, Arlington, Virginia, USA; Prof. Heidi Diggelmann, President of the Swiss National Science Foundation, Lausanne; Rainer Gerold, Director XII B.I – Life Sciences and Quality of Life, Commission of the European Communities, Brussels, Belgium; Prof. Michael Gibbons, Secretary General of the Association of Commonwealth Universities, London, United Kingdom; Prof. Bernard Giovannini, Département de Physique de la Matière Condensée, University of Geneva; Kirsten Hollaender, M.A., Research Institute for Sociology, University of Cologne, Germany; Sunita Kapila, IDRC Branch Office, Nairobi, Kenya; Prof. Ruth Kaufmann-Hayoz, Interdisciplinary Center for General Ecology, University of Berne; PD Gertrude Hirsch Hadorn, President of the Swiss Academic Society for Environmental Research and Ecology, Zurich; Prof. Pieter Leroy, University of Nijmegen, The Netherlands; Mag. Marie Céline Loibl, Austrian Institute for Applied Ecology, Vienna, Austria; Prof. Hansjürg Mey, President of the Swiss Federal Commission for Universities of Applied Sciences, Berne; Prof. Jürgen Mittelstrass, University of Constance, Germany; Sybille Oetliker, Cash, Zurich; Jan-Eric Sundgren, President of Chalmers University, Göteborg, Sweden; Dr. Johannes R. Randegger, Head of Infrastructure, Basel Works, Novartis Services Ltd, and member of the Swiss Parliament, Basel; Dr. Michel Roux, Swiss Federal Institute for Forest,

Snow and Landscape Research, Birmensdorf; Prof. Julie Thompson Klein, Wayne State University, Detroit, Michigan, USA; Prof. Albert Waldvogel, Vice President Research, Swiss Federal Institute of Technology, Zurich; Prof. Jörg Winistörfer, Vice Chancellor University of Lausanne; Betty Zucker, Head of the Department for Corporate Development and member of the Executive Management, Gottlieb Duttweiler Institute (GDI), Rüschlikon.

Appendix B

References

Note: All references to texts of the pre-conference Workbooks (I and II) are in the List of Contributors (Appendix D).

AEIDL (1997): Innovation and Rural Development. Brussels: European Rural Observatory.

Allen, P.M. (1997): Cities and Regions as Self-Organizing Systems: Models of Complexity. Reading: Gordon and Breach.

Amsler-Delafosse, S. (1998): Eau et Santé: Le Cas de Trois Quartiers Urbains à N'Djaména (Tchad). Research Report. N'Djaména, Chad: Swiss Tropical Institute.

Arber, W. (ed.) (1993): Inter- und Transdisziplinarität: Warum? – Wie? (Inter- et Transdisciplinarité, Pourquoi? – Comment?). Bern: Haupt.

Bachimon, P. (1998): Mobilization Communautaire pour la Distribution et la Protection de L'eau Potable dans un Environnement Urbain Défavorisé au Tchad et au Sénégal. Research Report. Paris: Pseau. Available at <http://home.worldnet.fr/~alainmh/>.

Beck, U., Giddens, A. and Lash, S., eds. (1994): Reflexive Modernization. Politics, Tradition and Aesthetics in the Modern Social Order. Stanford: Stanford University Press.

Becker, E. and Jahn, T., eds. (1999): Sustainability and the Social Sciences. London: Zed Books.

Becker, E. et al. (1997): Sustainability: A Cross-Disciplinary Concept for Social Transformations. UNESCO-MOST-Paper No.6. Paris: UNESCO.

BMWV (Bundesministerium für Wissenschaft und Verkehr) (1998): Szenarien der Kulturlandschaft. Forschungsschwerpunkt Kulturlandschaft No. 5. Vienna.

BMWV (Bundesministerium für Wissenschaft und Verkehr) (2000): Homepage of the Cultural Landscape Research Focus <http://www.klf.at>.

Brander, B., Hirsch, M., Meier-Dallach, H.P., Sauvain, P. and Stalder, U. (1995): Skitourismus. Von der Vergangenheit zum Potential der Zukunft. Zürich: Rüegger.

Britt, F. (2000): Social Insurance and Distribution of Health Care Services in Switzerland. In: Dialog Ethik (ed.) Reader for the Mutual Learning Session on Health Costs and Benefits. Zurich: Dialog Ethik.

Brundtland, G.H. (1987): Our Common Future. Report of the World Commission on Environment and Development. Oxford-New York: Oxford University Press.

Brunswik, E. (1952): The Conceptual Framework of Psychology. Chicago: University of Chicago Press.

D'Ambrosio, U. (1997): Universities and Transdisciplinarity. The Role of Universities in Modern Society. Annexe au Document de Synthèse Ciret-Unesco, Congrès de Locarno. 30 April–2 May 1997. Available at <http://perso.club-internet.fr/nicol/ciret/rechnom/rech.htm>.

Defila, R. and Di Giulio, A. (1999): Evaluating Transdisciplinary Research. PANORAMA (1), 1–28. Special issue on Evaluation.

Deutscher Bundestag (1999): Forschungs- und Technologiepolitik für eine Nachhaltige Entwicklung. Drucksache. 14/571/18.03.99.

Doubeck, C. and Zanetti, G. (1999): Siedlungsstruktur und öffentliche Haushalte. ÖROK-Schriftenreihe No. 143.

Doublier, G. and Dobingar. A. (1998): Table Ronde pour la Mise en Place d'un Plan D'action de Gestion Durable des Déchets Solides et de D'assainissement de la Ville de Djaména. Report. Municipality of N'Djaména, Chad.

Edvinsson, L. and Malone, M.S. (1997): Intellectual Capital. New York: Harper Collins.

Emmelin, L (1973): Report on Environmental Education at University Level. Paris. OECD.

Enda Graf (1993): La Resource Jumaine, Avenir de Terroirs: Recherches Paysannes au Sénégal. Paris: Kathala.

Encyclopedia of Life Support Systems (2000): Prospectus for Section 6.49: Unity of Knowledge (in Transdisciplinary Research for Sustainability). Oxford, UK: EOLSS Publishers.

Epstein, C. and Scarlet, T. (1975): The Ideal Marriage between the Economist's Macroapproach and the Social Anthropologist's Microapproach in Development Studies. Economic Development and Cultural Change (24, 1), 29–46.

Ernst, R. (1999): Mein Traum einer Idealen ETH. Abschiedsvorlesung. Kleine Schriften No. 39. 21.5.1999. Zurich: ETH.

Garreau, J. (1991): Edge City: Life on the New Frontier. New York: Doubleday.

Gerold, R. "Transdisciplinarity: From Concept to Reality." Remarks at Press Conference of the International Transdisciplinarity Conference. Zurich, Switzerland, 1 March 2000.

Gibbons, M., et al. (1994): The New Production of Knowledge: The Dynamics of Science and Research in Contemporary Societies. London, Thousand Oaks. New Delhi: Sage.

Giovannini, B. and Revéret, J. P. (1999): Definition and Practice of Transdisciplinarity for Sustainable Development: Concept, Methodologies and Examples. Preprint. Geneva.

Goldemberg, J., et al. (1985): An End-Use Oriented Global Energy Strategy. Annual Review of Energy (10), 613–88.

Häberli, R. (1995): Transdisciplinary Exchange within the SPP Environment. PANORAMA (5), 6–13.

Häberli, R. and Grossenbacher-Mansuy, W. (1998): Transdisziplinarität zwischen Förderung und Überforderung. Erkenntnisse aus dem SPP Umwelt. (Transdisciplinarity Between Granting and Overdemanding, Insights from the SPP Environment). GAIA (7,3), 196–213.

Häberli, R. et al. (2000): Transdisciplinarity: Joint Problem-Solving among Science, Technology and Society. Dialogue Sessions and Idea Market, Workbook I; Contributions to the Dialogue Sessions and Idea Market of the International Transdisciplinarity 2000 Conference. Zürich: Haffmans Sachbuch Verlag AG.

Hare, A.P. (1996): Handbook of Small Group Research, 2nd ed. New York: Free Press.

Heise-Verlag Online 2000. Telepolis.

Hofer, K. and Stalder, U. (2000): Regionale Produktorganisationen als Transformatoren des Bedürfnisfeldes Ernährung in Richtung Nachhaltigkeit? In: Georgraphica Bernensia, Liebefeld Bern: Lang Druck AG.

Hofstetter, P. (1998): Perspectives in Life Cycle Assessment. Boston: Kluwer.

Holländer, K. and Friedrichs, J. (1997): Theorien für die Praxis – Praxis für die Theorie. Theoretische Potentiale und praktische Probleme der Integration im Förderschwerpunkt "Stadtökologie". In: TA-Datenbank-Nachrichten Jg. 6, Nr. 3/4, 45–49.

Holt, E. (1997): Green Pricing. Resource Guide. Maine: The Regulatory Assistance Project.

Holzhey, H., et al, Eds. (1974): Interdisziplinär (Interdisciplinary). Ringvorlesung der Eidgenössischen Technischen Hochschule Zürich. Basel: Schwabe.

Holzinger, E. (2000): Räumliche Integration sozialer Gruppen. Incentive Paper for the Austrian Space Development Concept (ÖRK) 2001. Vienna: ÖIR.

Interdisciplinarity: Problems of Teaching and Research in Universities. (1972), Paris: OECD.

Horibe, F. (1999): Managing Knowledge Workers. Wiley, Etobicoke.

IUCN and World Bank (1997): Large Dams. Learning from the Past, Looking at the Future. Workshop Proceedings. April in Gland, Switzerland.

Jansen, J.L.A., et al. (1998): STD Vision 2040–1998, Technology, Key to Sustainable Prosperity. The Hague: H. ten Hagen & Stam, December. (In Dutch and English. ISBN 90-71694-86-0).

Jansen, J. L.A. (1993): Towards a Sustainable Future: En Route with Technology! In: The Environment: Towards a Sustainable Future, pp. 497–523. Dordrecht, Boston, London: Kluwer.

Jantsch, E. (1972): Towards Interdisciplinarity and Transdisciplinarity in Education and Innovation. In: Interdisciplinarity. Problems of Teaching and Research in Universities, 97–121. Paris: OECD.

Jungbluth, N., Tietje, O. and Scholz, R.W. (2000): Food Purchases: Impacts from the Consumers' Point of View Investigated with a Modular LCA. In: International Journal of Life Cycle Assessment (5,3), 134–142. <http://www.uns.umnw.ethz.ch/~jungblu/publication.html>.

Kauffman S. (1993): The Origins of Order: Self-Organization and Selection in Evolution. New York, Oxford University Press

Klein, J.T. (1996): Crossing Boundaries: Knowledge, Disciplinarities, and Interdisciplinarities. Charlottesville: University Press of Virginia.

Korczak, D. (2000): The Rationing Debate in Germany. In: Dialog Ethik. Reader for the Mutual Learning Session on Health Costs and Benefits. Zurich: Dialog Ethik.

Kötter, R. and Balsiger, Ph. (1997): Vom Wert Terminologischer Differenzierung: Rhetorik und Wissenschaftliche Sprache bezüglich Formen disziplinenübergreifender Wissenschaftspraktiken (Multidisziplinarität, Interdisziplinarität, Transdisciplinarität). Lecture at the First Workshop of the Forum Transdisziplinärer Forschungsprozess, March 18. Bern

Lattanzi, M. (1998): Transdisciplinarity. Stimulating Synergies, Integrating Knowledge. Division of Philosophy and Ethics, UNESCO. <http://www.unesco.org/philosophy/transdisciplinarity/index.html>.

Levaque L. (1998): Etude Qualitative Portant sur les Problèmes de Santé des Enfants des Rues de N'Djaména et Leur Recours aux Soins. Research Report. N'Djaména, Chad: Swiss Tropical Institute

Lipton, M. (1970): Interdisciplinary Studies in Less Developed Countries. The Journal of Development Studies (7, 1), 5–18.

Loibl, M.C. and Lechner, R. (1997): Interdisziplinäre Arbeitsmethoden und Umsetzungsziele in der Österreichischen Kulturlandschaftsforschung. Vienna: Federal Ministry of Science and Transport.

Loskow, P.L. (1998): Electricity Sectors in Transition. The Energy Journal, (19, 2), 25–54.

Luhmann, N. (1990): Ökologische Kommunikation. 3. Aufl. Westdeutscher Verlag Opladen.

Lukesch, R. (1999): Patterns of Sustainability. Module for the Integrated Training Program in the Research Project: Amazonia 21 – Operational Features of Managing Sustainable Development in Amazonian Countries. Unpublished Document, available on the homepage <www.amazonia21.org>.

Malik, F. (1992): Strategie des Managements Komplexer Systeme. Bern, Stuttgart, Wien: Verlag Paul Haupt.

Markard, J. and Truffer, B. (1999): Der lange Weg zu einem Euro-Label für Strom. In: Energiewirtschaftliche Tagesfragen. Zeitschrift für Energiewirtschaft, Recht, Technik und Umwelt (49, 11), 724–729.

Markl, H. (1999): Grundlagenforschung und Anwendungspraxis, Erfolg der Wechselwirkung. Wirtschaft & Wissenschaft, 1: 24–37.

Minsch, J., et al. (1998): Institutionelle Reformen für eine Politik der Nachhaltigkeit. Berlin u.a: Springer.

Mittelstrass, J. (1996): Stichwort Interdisziplinaritat. Mit einem anschliessenden Werkstattgespräch. Basel: Europainstitut an der Universität Basel.

MUT, Journal of the Styrian Chamber of Commerce (1998): City Lights – verlöscht die Stadt in der Peripherie? (41). Graz.

Nadjitolnan, O. (1999): Connaissances, Attitudes et Pratiques à N'Djaména en Matière de Paludisme et de Moustiquaires Imprégnées d'Insecticide. Research Report. N'Djaména, Chad Swiss Tropical Institute.

N'Diaye, M. (1999): Recherche Populaire. In: Environnement Urbain – Recherche et Action dans les Pays en Développement, Ed. Bolay, J.C., et al., pp. 23–28. Basel: Birkhäuser.

Nicolescu, B. (1997): Talk at the International Congress "Universities' Responsibilities to Society." International Association of Universities. Bangkok, Thailand Chulalongkorn University. November 12–14. <http://perso.club-internet.fr/nicol/ciret/bulletin/b12/b12c8.htm>.

Nicolis, G. and Prigogine, I. (1989): Exploring Complexity. New York: Freeman.

Nodjiadjim, A.L. and Wyss, K. (1999): Recours aux Soins des Enfants de la Rue. N'Djaména. Bulletin Medicus Mundi Switzerland (75), 16–17.

Nonaka, I., Takeuchi, H. (1995): The Knowledge-Creating Company. New York: Oxford University Press.

Nowotny, H. (1997): Transdisziplinäre Wissensproduktion – Eine Antwort auf die Wissensexplosion? In: Wissenschaft als Kultur. Österreichs Beitrag zur Moderne. Ed. F. Stadler, pp. 177–195. Vienna.

Ökologie-Institut: Ökoplan Weiz. (1996): Vienna. With continuous updates (Ökoplan 2000).

Organizing Committee Transdisciplinarity Conference (1999): Third Announcement International Transdisciplinarity Conference. February 27, 28, 29 and March 1, 2000 – Joint Problem Solving among Science, Technology and Society. Zurich.

ÖROK (Österreichische Raumordnungskonferenz) (1995): Möglichkeiten und Grenzen integrierter Bodenpolitik in Osterreich, verfasst vom KDZ (Kommunalwissenschaftliches Dokumentationszentrum) unter der Leitung von Georg Schadt.

Pawley, M.(2000): Architektur auf den Wellen des urbanen Niedergangs. Contribution on the homepage of the Heise-Verlag (Translation into German by Florian Rötzer). <http://www.heise.de/tp/deutsch/kolumnen/paw/2153/1.htm>.

Probst, G. Raub, S., and Romhard, K. (1999): Wissen Managen. 3. Aufl. Wiesbaden Gabler.

Program Management for SPPE, eds. (1995): Transdisciplinarity. PANORAMA (5), 1–72.

Rosenfield, P L. (1992): The Potential of Transdisciplinary Research for Sustaining and Extending Linkages between the Health and Social Sciences. Social Science and Medicine (35, 11), 1343–1356.

Scholz, R. W. and Tietje , O. (in prep.). Integrating Knowledge with Case Studies. London, Thousand Oaks, New Delhi: Sage.

Scholz, R. W. (1987): Cognitive Strategies in Stochastic Thinking. Dordrecht: Reidel.

Scholz, R.W., Bösch, S., Mieg, H.A., and Stünzi, J., eds. (1998): Region Klettgau – Verantwortungsvoller Umgang mit Boden. [Klettgau Region – responsible use of soil]. Zürich: Rüegger.

Scholz, R. W. (1999): "Mutual Learning" und probabilistischer Funktionalismus – Was Hochschule und Gesellschaft voneinander und von Egon Brunswik lernen können. ["Mutual Learning" and Probabilistic Functionalism – What University and Society can Learn from Each Other and from Egon Brunswik]. UNS-Working Paper, 21. ETH Zürich: Umweltnatur- und Umweltsozialwissenschaften.

Scholz, R. W., et al., eds. (1999): Nachhaltige Regionalentwicklung: Chancen der Region Klettgau. [Sustainable Regional Development: New Opportunities in the Klettgau Region] Zürich: Rüegger.

Scholz, R.W., et al., eds. (2000): Transdisciplinarity: Joint Problem-Solving among Science, Technology and Society. Mutual Learning Sessions, Workbook II; Contributions to the Mutual Learning Sessions of the International Transdisciplinarity 2000 Conference. Zürich: Haffmans Sachbuch Verlag AG.

Schremmer, Ch. (1999): Beitrag zum Entwicklungsprogramm Oststeiermark: Chancen und Wege einer nachhaltigen Siedlungsentwicklung. Expert paper for the regional management Oststeiermark. Wien: ÖIR.

Smit, B., et al. (1998): Beitrag zum Entwicklungsprogramm Oststeiermark: Chancen und Wege einer Nachhaltigen Siedlungsentwicklung. Expert Paper for the Regional Management Oststeiermark. Wien: ÖIR.

Smoliner, C., et al. (1998): Umweltwissenschaft im öffentlichen Auftrag; Vom Konzept zur Forschungspraxis. Vienna: Federal Ministry of Science and Transport.

Stadt Zürich (1999): Hochbaudepartement der Stadt Zürich: Kooperative Entwicklungsplanung Zürich West. Synthesebericht der Stadt Zürich und der Mitwirkenden Grundeigentümer. März.

Stalder, U. (1995): Skitourismus. Von der Vergangenheit zum Potential der Zukunft. Chur, Zürich: Rüegger.

Susskind, L., McKearnen, S. and Thomas-Larmer J., eds. (1999): The Consensus Building Handbook: A Comprehensive Guide to Reaching Agreement. London, Thousand Oaks, New Delhi: Sage.

Swiss National Science Foundation, eds. (1995): Plan of Execution for the Swiss Priority Program (SPP) Environment, 1996–1999.

Truffer, B. (1999): Green Electricity Products and the Sustainable Use of Water Resources Forum of the UNESCO International School of Science for Peace on Water Security in the Third Millennium. Mediterranean Countries as a Case. Centro Volta, Como: Landau Network .

Truffer, B., et al. (1998): 'Ökostrom': Transdisziplinarität auf der Werkbank. GAIA (7, 1), 26–35.

Villiger, A., Wüstenhagen, R., and Meyer, A. (2000): Jenseits der Öko-Nische. Basel: Birkhäuser Verlag.

Wandel, J (1998): Agroecosystem Health – Analysis and Assessment. University of Guelph, Canada.

Weaver, P. et al. (2000): Sustainable Technology Development. Sheffield UK: Greenleaf Publishing. ISBN 1-874719-09-8.

Willke, H. (1998): Systemisches Wissensmanagement. Stuttgart: UTB.

Wissenschaftlicher Beirat der Bundesregierung "Globale Umweltveränderungen" (WBGU) (1996): Welt im Wandel – Herausforderung für die deutsche Wissenschaft, Jahresgutachten 1996, Berlin: Springer.

WKSt (Wirtschaftskammer Steiermark) (2000): Kern Gesund – Gemeinde Gesund. Eine Wirtschaftskammerinitiative zur Erhaltung gewachsener Steirischer Zentren. Internet Site <http://www.ortskern.at/inhalt.htm>.

Woodhill, J. and Röling, N. G. (1998): The Second Wing of the Eagle: The Human Dimension in Learning Our Way to More Sustainable Futures. In: Facilitating Sustainable Agriculture, Ed. N. G. Riling and M.A. Wagemakers, pp. 46–69. Cambridge: Cambridge University Press.

Wynne, B. (1996): May the Sheep Safely Graze: A Reflective View on Expert-Lay Knowledge. In: Szerszynski. Risk Environment and Modernity, Ed. Lash, S., Wynne, B. & B. London, Thousand Oaks, New Delhi: Sage.

Wyss, K. (1999): Recherche – Action – Formation à N'Djaména. In: Environnement Urbain – Recherche et Action dans Les Pays en Développement. Ed. Bolay, J.C, et al., pp. 137–42. Basel: Birkhäuser.

Yémadji, N. et al. (1999): La Recherche – Action – Formation Appliquée à la Gestion d'un Espace Urbain Défavorisé. In: Environnement Urbain – Recherche et Action dans Les Pays en Développement. Ed. Bolay, J.C. et al, pp. 143-154. Basel: Birkhäuser.

Zibell, B. and Gürtler-Berger, E. (1997): Stadt im Umbruch: ChaosStadt? Zurich, ETH.

Ziman, J. (1994): Prometheus Bound. Science in a Dynamic Steady State. Cambridge: Cambridge University Press.

Appendix C

List of Participants

Abbasi Shahid, Pondicherry University, Pondicherry, India, prof_abbasi@vsnl.com; **Abou-Khaled** Omar, Ecole d'Ingénieurs de Fribourg, Fribourg, aboukhal@di.epfl.ch; **Abplanalp** Peter, University of Applied Sciences, Solothurn, peter.abplanalp@fhso.ch; **Achiaa** Amma, Our affairs in communities, Accra, Ghana; **Adégnika** Félix, Alter Ego/PDM, Cotonou, Bénin, pdm@bow.intnet.bj; **Adey** Bryan, Swiss Federal Institute of Technology, EPF-Lausanne, Bryan.Adey@epfl.ch; **Adi** Alpheus Bongo C., Restral Consulting Limited, Victoria Island Lagos, Nigeria, restral@infoweb.abs.net; **Aebersold** Michael, Bundesamt für Energie, Bern, michael.aebersold@bfe.admin.ch; **Aenis** Thomas, Humbolduniversität Berlin, Landwirtschaftlich-gärtnerische Fakultät, Berlin, Germany, thomas.aenis@agrar.hu-berlin.de; **Agger** Peder, Roskilde, Denmark, pa@teksam.ruc.dk; **Aiking** Harry, Vrije Universiteit, Amsterdam, The Netherlands, harry.aiking@ivm.vu.nl; **Alabor** Kurt, Amt für Umweltschutz, St. Gallen; **Albrecht** Stephan, University of Hamburg, FSP BIOGUM, Husum, Germany, alwold5@aol.com; **Aliò i Torres** Maria Àngels, University of Barcelona, Spain, angelsa@trivium.gh.ub.es; **Amarawickrama** Tito, Kandy, Sri Lanka; **Ammann** Paul, Gesundheits- und Fürsorgedirektion Kt. Bern, paul.ammann@gef.be.ch; **Arbenz** Benno, Zollikon, arbenz@ifi.unizh.ch; **Ashida** Makiko, SAM Sustainability Group, Zollikon, makiko@sam-group.com

Bachmann Carine, University of Geneva, carine.bachmann@politic.unige.ch; **Bachmann** Lukas, Horten-Zentrum, Zurich, Lucas.bachmann@dim.urz.ch; **Bachmann** Reinhard, ETH Zurich, bachmann@ifap.bepr.ethz.ch; **Baer** Alec, Belp; **Bährer** Sabine, Psychiatric University Hospital, Basel, sabine.baehrer@datacomm.ch; **Baitsch** Christof, IAP Zurich, chbaitsch@iap.psy.ch; **Bal** Lucinia, Open Society Institute, Budapest, Hungary, lbal@osi.hu; **Balsiger** Philipp W., University of Erlangen-Nuremberg, Erlangen, Germany, pebalsig@phil.uni-erlangen.de; **Bambara** Emanuela, Istituto di Filosofia Università di Messina, Italy, magazu@dsme01.messina.infm.it; **Banskota** Pradeep, APEC, Biratnaga, Nepal, spacecom@ccsl.com.np; **Baracchi** Claudio, ABB Immobilien AG, Baden, claudio.baracchi@ch.abb.com; **Bastian** C., DOW Europe, Horgen, cbastian@dow.com; **Batzer** Martin, Novartis Pharma Schweiz AG, Bern; **Baud** Roger, ETH Zurich, baud@sl.ethz.ch; **Bauer** Werner, Kollegium für Hausarztmedizin, Küssnacht, werner.bauer@hin.ch; **Baumann** Barbara, Institute of Biotechnology, Zurich, baumann@biotech.biol.ethz.ch; **Baumann** Maura, ETH, Swiss Federal Institute of Technology Zurich, baumann@sl.ethz.ch; **Baumann** Max, University of

Zurich/Interdisziplinäres Institut für Ethik im Gesundheitswesen, Zurich, baujus@swissonline.ch; **Baumann-Hölzle** Ruth, Wolfshausen; **Baumberger** Ernst, Alpenregion Brienz-Meiringen-Hasliberg, Meiringen; **Baumgartner** Rudolf, ETH Zurich/NADEL, Zurich, baumgartner@nadel.ethz.ch; **Baumgartner** Thomas, Swiss Federal Institute of Technology (ETHZ), Natural & Social Science Interface (UNS), Zurich, baumgartner@uns.umnw.ethz.ch; **Baur** Bruno, University of Basel, Bruno.Baur@unibas.ch; **Bayili** Paul Pérré, Alter Ego/Ville de Ouagadougou, Burkina Faso, bayili@fasonet.bf; **Bearth** Thomas, University of Zurich, Thomas_Bearth@compuserve.com; **Beck** Almut, GAIA, Zurich; **Beck** Hans, University of Neuchâtel, Hans.Beck@iph.unine.ch; **Becker** Egon, Institut für sozial-ökologische Forschung, Frankfurt am Main, Germany, E.Becker@em.uni-frankfurt.de; **Beeler** Rene, Zürcher Kantonalbank, Zurich, rene.beeler@zkb.ch; **Begusch-Pfefferkorn** Karolina, Programme Coordination Austrian Landscape Research, Vienna, Austria, karolina.begusch@klf.at; **Behringer** Jeannette, Swiss Federal Institute for Environmental Science and Technology, Dübendorf, jeannette.behringer@eawag.ch; **Bello** Bugallo Pastora, University of Santiago de Compostela, Espana, eqpmbb@usc.es; **Bellucci** Sergio, Swiss Science and Technology Council, Bern, sergio.bellucci@swr. admin.ch; **Berg** Hansen Inge, Danish Research Agency, Copenhagen, Denmark, berg@forsk.dk; **Berg** Marco, Erdöl Vereinigung, Zurich, berg@swissoil.ch; **Bernhard** Hans-Peter, Novartis Services AG, Basel, hans-peter.bernhard@sn.Novartis.com; **Bertholet** Miriam, Freiburg i. Br., Germany, berthole@psychologie.uni-freiburg.de; **Bertilsson** Britt Marie, Foundation for Strategic Environmental Research, Stockholm, Sweden, bm.bertilsson@mistra-research.se; **Bertschi** Annen Judith, Illnau, annen@dataway.ch; **Bianculli** Claudio, ZAB, Bazenheid; **Biber-Klemm** Susette, University of Basel, Susette.Biber-Klemm@unibas.ch; **Bierens de Haan** Camille, Institut universitaire Kurt Bösch, Sion, camille.bdehaan@ikb.vsnet.ch; **Bieri** Françoise, Bio Tech Forum, BICS, Basel; **Bilang** Roland, InterNutrition, Zurich, rb@-internutrition.ch; **Bill** Alain, ABB Corporate Reseach Ltd., Baden, alain.bill@ch.abb.com; **Binder** Andres, Novartis, Basel, andres.binder@cp.novartis.com; **Bingzhang** Xue, ABB, Energy & Global Change Department, Baden-Daettwil, bingzhang@ch.abb.com; **Blatter** Roland, Amt für Umweltschutz, Liestal, Roland.Blatter@bud.bl.ch; **Blowers** Andrew, Open University, Milton Keynes, UK, BlowersA@csd.bedfordshire.gov.uk; **Bocharnikov** Vladimir N., Pacific Institute of Geography, Vladivostok, Russia, sergeikr@online.vladivostok.ru; **Boettcher** Sabine, University of St. Gallen, sabine.boettcher@unisg.ch; **Bolay** Jean-Claude, Swiss Federal Institute of Technology/IREC, Lausanne, Jean-Claude.Bolay@epfl.ch; **Bold** Mario, University of Zurich, mbold@geo.unizh.ch; **Bollmann** Kurt, Bird Life Switzerland, Zurich, kurt.bollmann@birdlife.ch; **Borner Schweizer** Sybille, Winterthur, s.borner@freesurf.ch; **Böschen** Stefan, Institut für Soziologie, Augsburg, Germany, stefan.boeschen@wiso.uni-augsburg.de;

Braun Richard, Bio-Link, Worb, rdbraun@bluewin.ch; **Brenner** Joseph, OEUF, Les Diablerets, jebrenner@compuserve.com; **Britt** Fritz, Bundesamt für Sozialversicherung, Bern; **Broggi** Mario F., Eidg. Forschungsanstalt für Wald, Schnee und Landschaft, Birmensdorf; **Browne** Neil, Bowling Green State University, Bowling Green, USA, nbrown2@cba.bgsu.edu; **Brühwiler** Barbara, Muri; **Brunner** Andreas, ETH Zurich, brunnand@student.ethz.ch; **Brunner** Hans Heinrich, Bern; **Brunner** Ursula, Zurich, u.brunner@aebsb-anwaelte.ch; **Brunold** Sylvia, LBL, Eschlikon, eza@lbl.ch; **Brunschwig Graf** Martine, Direction, Genève, martine.brunschwiggraf@etat.ge.ch; **Buchecker** Matthias, Swiss Federal Institute for Forest, Snow and Landscape Research, Birmensdorf, matthias.buchecker@wsl.ch; **Büchel** Dominik, Locher, Brauchbar & Partner, Basel, Buechel@lbp.ch; **Bucheli** Erika, Swiss National Science Foundation, Bern, ebucheli@snf.ch; **Bucher** Hans-Ulrich, Neonatologie USZ, Zurich; **Büchi** Hansjürg, Zurich, buh@sunweb.ch; **Büchi** Martin, Horten-Zentrum, Zurich; **Buchinger** Eva, Austrian Research Centers Seibersdorf (ARCS), Seibersdorf, Austria, eva.buchinger@arcs.ac.at; **Buechele** Bruno, University of Karlsruhe (TH), Germany, bruno.buechele@bau-verm.uni-karlsruhe.de; **Büechi** Martin, Swiss Federal Office of Public Health, Bern, martin.buechi@bag.admin.ch; **Bunge** Rainer, Eberhard Recycling AG, Kloten, rainer.bunge@eberhard.ch; **Burgenmeier** Beat, Université de Genève, Faculté des S.E.S., Genève, Beat.Burgenmeier@ses.unige.ch; **Burger** Max M., Friedrich Miescher Institut, Basel, burger@fmi.ch; **Burger** Paul, University of Basel, paul.burger@unibas.ch; **Buschor** Ernst, Staatskanzlei des Kantons Zurich; **Bütschi** Danielle, Swiss Science and Technology Council (SSTC), Bern, danielle.buetschi@swr.admin.ch

Caetano João Carlos, Universidade Aberta, Lisboa, Portugal, joaocaetano@mail.telepac.pt; **Camartin** Iso, Zurich; **Campagna** Maurice, Chief Technology Officer ABB Alstom Power Ltd., Brussels, maurice.campagna@ch.abb.com; **Caravita** Silvia, Ist. di Psicologia of Nat. Res. Council (CNR), Roma, Italy, caravita@ip.rm.cnr.it; **Carter** Claudia, University of Cambridge, UK, cec29@cam.ac.uk; **Cassina** Enrico, Sieber Cassina + Partner AG, Olten, cassina@scpag.ch; **Catani** Reto, Teachers Training College, Spiez, mkuebler@bluemail.ch; **Caviola** Hugo, University of Basel, Liestal, caviolah@datacomm.ch; **Charlesworth** Mark, Keele University, Keele, Staffordshire, UK, m.e.charlesworth@pol.keele.ac.uk; **Chevallaz** Roger, reflecta ag, Bern, chevallaz@reflecta.ch; **Chikhi-Jans** Gabriele, ASL, Genève, asl@worldcom.ch; **Christ** Urs, SNSF, Bern, uchrist@snf.ch; **Cissé** Guéladio, CSRS, Abidion, Côte d'Ivoire, csrs-urbain@globeaccesse.net; **Colwell** Rita, US National Science Foundation, Arlington, USA; **Cornelis** Gustaaf, SCK-CEN, Mol, Belgium, gcorneli@sckcen.be; **Corneloup** Thierry, Genève, thierrycorneloup@voila.fr; **Cottier** Thomas, Institut für Europa- und Wirtschaftsvölkerrecht, Bern, thomas.cottier@iew.unibe.ch; **Crausaz** Németh Roselyne, Forum

Engelberg, Fribourg, th.wolf@bluewin.ch; **Craye** Matthieu, University of Antwerp (UFSIA), Antwerpen, Belgium, matthieu.craye@ufsia.ac.be; **Cslovjecsek** Markus, Niederwil, m.cs@bigfoot.com; **Cueni** Thomas, Geschäftsführer Interpharma, Basel; **Cugini** Carla, Zurich, carla.cugini@access.unizh.ch

Dahinden Urs, University of Zurich, dahind@ipmz.unizh.ch; **Dahme** Miriam, Zürcher Innovationszentrum, University of Zurich, dahme@ziwig.unizh.ch; **Darmaratna** U.G.N.M.S., Kandy, Sri Lanka; **de Buman** Anne-Marie, Secrétariat d' Etat à la Science et à la Recherche, Bern, anne-marie.debuman@gwf.admin.ch; **De Mello** Maria F., University of São Paulo, Brasil, cetrans@futuro.usp.br; **De Roulet** Daniel, Science et Cité, Bern, info@science-et-cite.ch; **de Wit** Bert, Advisory Council for Research on Nature and the Environment, Rijswijk, The Netherlands, rmno@xs4all.nl; **Decker** Michael, Europäische Akademie zur Erforschung von Folgen wissenschaftlich-technischer Entwicklungen, Bad Neuenahr-Ahrweiler, Germany, michael.decker@dlr.de; **Defila** Rico, University of Bern, defila@ikaoe.unibe.ch; **Demyanenko** Valery, Engineering & Technology Ins., Cherkassy, Ukraine, is2000@chiti.uch.net; **Denz** Martin D., Ärztliche Direktion USZ, Zurich, martin.denz@vdi.usz.ch; **Derks** Lucas, Nijmegen, The Netherlands, Mol.Derks@Hetnet.NL; **Di Giulio** Antonietta, University of Bern, digiulio@ikaoe.unibe.ch; **Dias Nunes** Paulo Augusto Lourenço, Free University of Amsterdam, The Netherlands, pnunes@econ.vu.nl; **Diggelmann** Heidi, Präsidentin SNF, Lausanne, Heidi.Diggelmann@chuv.hospvd.ch; **Döbel** Reinald, Institut für Soziologie der Westfälischen Wilhelms-Universität, Bad Schwalbach, Germany; **Dobingar** Allassembaye, Swiss Tropical Institute, Basel, adobingar@hotmail.com; **Dolt** Claudine, Basel, claudine.dolt@unibas.ch; **Dovbysheva** Tatjana, Belarussian State Politechnical Academy, Minsk, Belarus, tdovbysheva@bspa.unibel.by; **Drilling** Matthias, Höhere Fachschule für soziale Arbeit, Basel, mdrilling@hfsbb.unibas.ch; **Dürrenberger** Gregor, Swiss Federal Institute for Environmental Science and Technology, Dübendorf, gregor@ifh.ee.ethz.ch

Eberle Armin, Migros Genossenschaftsbund MGB, Zurich, Armin.Eberle@mgb.ch; **Eggermont** Gilbert, SCK.CEN, Brussels, Belgium, geggermo@sckcen.be; **Egli** Gustav, Stiftung Diakoniewerk, Zollikerberg; **Ehmayer** Cornelia, 17&4, Vienna, Austria, cornelia.ehmayer@magnet.at; **Eichholzer** Erika, University of Hamburg, St. Gallen, Erika.Eichholzer@excite.com; **Eigenmann** Kathrin, University of Basel, Röschenz, eigkat00@stud.unibas.ch; **Einsele** Arthur, Novartis Seeds, Basel, arthur.einsele@seeds.novartis.com; **Eisenstein** Robert A., National Science Foundation, Arlington, USA, reisenst@snf.gov; **Eisner** Manuel, ETH Zentrum, Zurich, eisner@soz.gess.ethz.ch; **Eliasson** Baldur, ABB Corporate Research Ltd, Baden-Daettwil; **Elmiger** Marc, St. Gallen, marc.elmiger@bluewin.ch; **Engelson** Boris, Genève; **Engquist** Anders, Vårdalstiftelsen, Stockholm, Sweden, anders.engquist@vardal.se; **Enz** Anita, Amt für Umwelt, Frauenfeld, anita.enz@kttg.ch; **Erbetta** Marc Auguste, Dir. Environnement Conseil, Yverdon, erbetta@bluewin.ch; **Ericson** Sven-olov, Vattenfall Utveckling AB, Stockholm, Sweden, svenolov.ericson@utveckling.vattenfall.se; **Ernst** Richard, Physikalische Chemie, Zurich; **Estermann** Rita, Luzern

Fahrni Hans-Peter, BUWAL, Bern, hanspeter.fahrni@buwal.admin.ch; **Fan** Diomande, University of Kassel, Alheim-Heinebach, Germany, DiomandeFan@aol.com; **Farago** Peter, Landert Farago Davatz & Partner, Zurich, pfarago@access.ch; **Fehr** Ernst, Emp. Wirtschaftsforschung, Zurich; **Fehr** Johannes, Collegium Helveticum, Zurich; **Felber** François, University of Neuchâtel, Francois.felber@bota.unine.ch; **Fenchel** Marcus, ETH, Zurich, fenchel@uns.umnw.ethz.ch; **Fernando** Jeewani, AGRO MASS, Dehiwala, Sri Lanka, jeewani@eureka.lk; **Filipov** Dmytro, National Technical University of Ukraine (KPI) (IASA), Kyiv, Ukraine, fildm@yahoo.com; **Fischer** Joachim, Horten-Zentrum, Zurich; **Fisch-Märki** Josy E., Bremgarten, fisch@spectraweb.ch; **Flamm** Michael, Nyon, michael.flamm@epfl.ch; **Flesch** Joachim, 3Sat, Mainz, Germany, joachim.flesch@gmx.de; **Flores** Fernandez, Instituto de Cooperacion Internacional, Madrid, España, manuel_flores@hotmail.com; **Flückiger** Federico, NDIT/FPIT, Bern, ffl@ndit.ch; **Flüeler** Thomas, Umweltrecherchen & -gutachten, Hausen AG, flueeler_urg@bluewin.ch; **Flury** Manuel, University of Bern, flury@ikaoe.unibe.ch; **Foerster** Ruth, University of Basel, ruth.foerster@unibas.ch; **Follath** Ferenc, University Hospital Zurich, dimscy@usz.unizh.ch; **Franke** Sassa, Choice Mobilitätsproviding GmbH, Berlin, Germany, franke@choice.de; **Frei** Hans-Peter, UBS AG, Zurich, hans-peter.frei@ubs.com; **Frei-Bischoff** Rudolf, Swiss Reinsurance Company, Zurich, Rudolf_FreiBischoff@swissre.com; **Freimüller** Pierre, appunto communications, Glattbrugg, appunto@smile.ch; **Freyer** Bernhard, Institute of Organic Farming, Vienna, Austria, bfreyer@edv1.boku.ac.at; **Frischknecht** Peter, Swiss Federal Institute of Technology, Zurich, frischknecht@umnw.ethz.ch; **Frischknecht** Ursula, Sargans; **Fritschi** Albert, ETH-Rat, Zurich, fritschi@ethrat.ch; **Frund** Vinciane, Swiss Federal Office for Public Health, Bern, vinciane.frund@bag.admin.ch; **Fry** Patricia Elizabeth, Swiss Federal Institute of Technology, Zurich, fry@wawona.ethz.ch; **Füglister** Stefan, Greenpeace Zurich, stefan.fueglister@ch.greenpeace.org; **Füssler** Jürg, Ernst Basler + Partner AG, Zurich, juerg.fuessler@ebp.ch

Gähwiler Manuela, ETH Zurich, Zollikon, mgaehwiler@student.ethz.ch; **Gallati** Justus, Kant. Amt für Umweltschutz, Luzern, gallati@die-luft.ch; **Ganguin** Jacques, Amt für Gewässerschutz und Abfallwirtschaft d. Kt. Bern, jacques.ganguin@bve.be.ch; **Gans** Werner, GAIA, Berlin, Germany; **Garcia** Marcos, University of St. Gallen, marcos.garcia@unisg.ch; **Gartmann** Samuel, Maag Holding,

Zurich; **Gautier** Laurent, Conservatoire et Jardin botaniques de la Ville de Genève, Chambéry, laurent.gautier@cjb.ville-ge.ch; **Gehler** Gaby, Zurich; **Geiger** Hans, Swiss Banking Institute, University of Zurich, geiger@isb.unkizh.ch; **Geissbühler-Greco** Isabella, Swiss National Science Foundation, Bern, Geissbuehler@sppe.ch; **Genske** Dieter D., Swiss Federal Institute of Technology, Lausanne, dieter.genske@epfl.ch; **Gerber** Alexander, Projektgruppe Kulturlandschaft Hohenlohe, Stuttgart, Germany, gerberal@uni-hohenheim.de; **Gerber** Beat, Tages-Anzeiger, Zurich; **Gerber** Mariette, INSERM, Montpellier, France, marietger@valdorel.fnclcc.fr; **Gerold** Rainer, Life Sciences and Quality of Life, Commission of the European Communities, Brussels, Belgium, Rainer.gerold@cec.eu.int; **Gersbach** Klaus, LIB Strickhof, Lindau; **Gerster** Richard, Gerster Development Consultants, Richterswil, rgerster@active.ch; **Geslin** Philippe, IMRA-SAD, Castanet-Tolosan, France, pgeslin@toulouse.inra.fr; **Gessler** Monika, ETH, Zurich, gessler@sl.ethz.ch; **Gessler** Rahel, Swiss National Science Foundation, Bern, gessler@sppe.ch; **Gessner** Wolfgang, Hochschule fur Wirtschaft, Olten, wolfgang.gessner@ikaoe.unibe.ch; **Gibbons** Michael, Secretary General Association of Commonwealth Universities, London, UK, secgen@acu.ac.uk; **Gilgen** Paul W., EMPA, St. Gallen, paul.gilgen@empa.ch; **Giovannini** Bernard, DP MC 24, Genève, giovanni@sc2a.unige.ch; **Girardi** Peter, Sozialmed. Organisation GmbH, Bregenz, Austria, peter.girardi@smo.at; **Gisin** Sandra, Zürcher Innovationszentrum, University of Zurich, Zurich, sgisin@zuv.unizh.ch; **Gisler** Priska, Swiss Federal Institute of Technology, Zurich, gisler@wiss.gess.ethz.ch; **Glauser** Christoph, MMS Media Monitoring Switzerland AG, Bern, glauser@mmsag.ch; **Gmünder** Felix K., Basler & Hofmann AG, Zurich, fgmuender@bhz.ch; **Gokalp** Iskender, CNRS-LCSR, Orleans, France, gokalp@cnrs-orleans.fr; **Goldbeck-Wood** Sandy, British Medical Journal, London, UK; **Goldschmid** Marcel, EPFL-CPD, Lausanne, marcel.goldschmid@epfl.ch; **Goorhuis** Henk, Universitäre Weiterbildung, Zurich, goorhuis@wb.unizh.ch; **Gotsch** Nikolaus, Swiss Federal Institute of Technology, Zurich, nikolaus.gotsch@iaw.agrl.ethz.ch; **Graf** Martin Anthony, Town of Illnau-Effretikon, Effretikon, graf@ilef.ch; **Grasmück** Dirk, Zurich, grasmueck@uns.umnw.ethz.ch; **Greutert** Mina, Zurich; **Grewal** Penny, Novartis, Basel, penny.grewalwilliams@group.novartis.com; **Gritsevitch** Inna, Center of Energy Efficiency, Moscow, Russia, cenef@glas.apc.org; **Gros** Henry, Innovation+Management, Winterthur, henry.gros@bluewin.ch; **Grossenbacher-Mansuy** Walter, Swiss National Science Foundation, Bern, grossenbacher@sppe.ch; **Grote** Gudela, Institut für Arbeitspsychologie, Zurich, grote@ifap.bepr.ethz.ch; **Grütter** Rolf, University of St. Gallen; **Guillemin** M., Lausanne, Michel.Guillemin@inst.hospvd.ch; **Güldenzoph** Wiebke, ETH Zentrum, Zurich, gueldenzoph@uns.umnw.ethz.ch; **Gutermann** Thomas, Zurich, t.gutermann@bluewin.ch; **Gutscher** Heinz, University of Zurich, gutscher@sozpsy.unizh.ch

Haab Markus, Bietenholz; **Häberli** Katharina, Direktion für Entwicklung und Zusammenarbeit DEZA, Bern, Katharina.Haeberli@deza.admin.ch; **Häberli** Rudolf, Swiss National Foundation, Bern, haeberli@sppe.ch; **Häberli-Banholzer** Martha, Bern; **Hadjieva** Mila, Bulgarian Academy of Sciences, Sofia, Bulgaria, hadjieva@bas.bg; **Haering** Barbara, econcept AG, Zurich, barbara.haering@ econcept.ch; **Häfeli** Ueli, University of Bern, haefeli@ikaoe.unibe.ch; **Hagmann** Alex, arte Wissenschaftsmagazin, Basel; **Hajdin** Rade, Swiss Federal Inst. of Technology, Lausanne, Rade.Hajdin@epfl.ch; **Hamm** Bernd, University of Trier, Germany, hamm@uni-trier.de; **Hänni** Heinz, Bundesamt für Landwirtschaft, Bern, heinz.haenni@blw.admin.ch; **Haribabu** E., University of Hyderabad, India, ehbss@uohyd.ernet.in; **Harms** Sylvia, EAWAG, Dübendorf, sharms@eawag.ch; **Hatz** Christine, University of Basel, christinehatz@datacomm.ch; **Hediger** Roland, Fernwärmeversorgung und KVA der Stadt Bern, roland.hediger@ bern.ch; **Heeb** Johannes, Seecon GmbH, Wollhusen, heebjohannes@pingnet.ch; **Held** Thomas, held planung und nachhaltigkeit, Zurich, held.th@bluewin.ch; **Hellmann-Grobe** Antje, Risiko-Dialog, Steinenbronn, Germany; **Henz** Alexander, Swiss Federal Institute of Technology, Auenstein, henz@arch.ethz.ch; **Hepperle** Erwin, Institut für Kulturtechnik, Zurich, hepperle@recht.gess.ethz.ch; **Hermanns Stengele** Rita, Swiss Federal Institute of Technology, Zurich, hermanns@igt.baug.ethz.ch; **Herren** Heinz, Gemeindeverwaltung Hasliberg, Hasliberg Goldern; **Herren** Madeleine, University of Zurich, mherren@ hist.unizh.ch; **Hertlein** Markus, IFOK, Institut für Organisationskommunikation, Bensheim, Germany, hertlein@IFOK.de; **Hertz** Jürg, Amt für Umwelt, Frauenfeld, juerg.hertz@auw.tg.ch; **Herzog** Ernst G., EGH Licensing, Basel, egherzog@ datacomm.ch; **Hess** Christian, Bezirksspital Affoltern a/A, ch.hess@freesurf.ch; **Hesske** Stefan, UNS-HAT-ETH Zurich, hesske@uns.umnw.ethz.ch; **Hess-Lüttich** Ernest W.B., Institut für Germanistik der Universität Bern, ernest.hess-luettich@germ.unibe.ch; **Heusser** Rolf, University of Zurich, rheusser@ ifspm.unizh.ch; **Hideto** Moritsuka, Central Research Institute of Electric Power Industry (CRIEPI), Yokosuka-City, Kanagawa-Prefecture, Japan, moritsuk@ criepi.denken.or.jp; **Hiess** Helmut, Rosinak & Partner, Vienna, Austria, sekretariat@rosinak.co.at; **Hilse** Jürgen, Kreissparkasse Göppingen, Germany, info@ksk-gp.de; **Hiltbrunner** Beat, Suva, Luzern, b_hiltbrunner@hotmail.com; **Hilty** Lorenz, FH Solothurn NW-Schweiz, Solothurn, lorenz.hilty@fhso.ch; **Hirsch Hadorn** Gertrude, President SAGUF, Zurich, hirsch@umnw.ethz.ch; **Hirsch** Jemma Madeleine, Geoprognos hirsch AG, Zurich, hirsch@ swissonline.ch; **Hirstein** Andreas, Bulletin SEV/VSE, Fehraltorf; **Hirt** Matthias, NDIT/FPIT, Bern, hirt@ndit.ch; **Hisschemoller** Matthijs, Institute for Environmental Studies, Amsterdam, The Netherlands, matthijs.hisschemoller@ivm.vu.nl; **Hofer** Andreas, Zurich, hofer@uns.umnw.ethz.ch; **Hoffelner** Wolfgang, RWH consult GmbH, Oberrohrdorf, rwh_whoffelner@compuserve.com; **Hoffmann**

Jörg, Galfingue, France, j.hoffmann@wanadoo.fr; **Hoffmann** Walter K.H., Transform Management Consulting, Kilchberg, whoffmanntmc@access.ch; **Hoffmann-Riem** Holger, Swiss Federal Institute of Technology, Schlieren, Hoffmann-Riem@ito.umnw.ethz.ch; **Högger** Rudolf, Stettlen; **Hohn** Barbara, Friedrich Miescher Institut, Basel; **Hollaender** Kirsten, University of Cologne, Köln, hollaender@wiso.uni-koeln.de; **Holliger** Christoph, University of Applied Sciences, Windisch, ch.holliger@fh-aargau.ch; **Holm** Patricia, EAWAG, Dübendorf, patricia.holm@eawag.ch; **Holzer** Andreas, Zürcher Kantonalbank, Zurich, andreas.holzer@zkb.ch; **Hölzle** Walter P., Warner Lambert (Schweiz) AG, Baar; **Höpflinger** François, Soziologisches Institut, Zurich, fhoepf@soziologie.unizh.ch; **Horisberger** Bruno, FMiG, St.Gallen, Evelyn.Reiter@fhwsg.ch; **Hroch** Millier Claude, Ecole Nationale du Genie Rural, Paris, France, millier@engref.fr; **Hroch** Nicole, University of Lüneburg, schaltegger@uni-lueneburg.de; **Hubschmid** Walter, Paul Scherrer Institut, Villingen PSI, walter.hubschmid@psi.ch; **Hugentobler** Margrit, Swiss Federal Institute of Technology, Zurich, hugentobler@arch.ethz.ch; **Huggenberger** Peter, University of Basel, peter.huggenberger@unibas.ch; **Hugi** Markus, Nagra, Wettingen, hugi@nagra.ch; **Hulmann** Hanspeter, Stadt Winterthur; **Hurni** Hans, Swiss Commission for Research Partnerships with Developing Countries (KFPE) of CASS, Bern, hurni@giub.unibe.ch; **Hürzeler** Beat, University of Bern, hurzeler@giub.unibe.ch; **Huter** Christoph, Entsorgung und Recycling Zurich, christoph.huter@erz.stzh.ch; **Hvid** Helge, Roskilde University, Roskilde, Denmark, hh@teksam.ruc.dk

Inwyler Charles, Zurich; **Iseli** Claudia, SRG SSR, Bern, claudia.iseli@sri.ch; **Jabbar** Mohammad, International Livestock Research Institute, Addis Ababa, Ethiopia, M.Jabbar@cgiar.org; **Jacquinet** Marc, Universidade Aberta, Lisboa, Portugal, mjacquinet@univ-ab.pt; **Jaeger** Jochen, Center of Technology Assessement in Baden-Württemberg, Stuttgart, Germany, jochen.jaeger@ta-akademie.de; **Jaeggi** Olivier, ECOFACT AG, Zurich, contact@ecofact.ch; **Jahn** Thomas, Institute for Social-ecological Research (ISOE), Frankfurt am Main, Germany, jahn@isoe.de; **Jans** Beat, Pro Natura, Basel, beat.jans@pronatura.ch; **Jansen** J.L.A., Sustainable Technology Development – Knowledge Dissemination and Anchoring, Delft, The Netherlands, jansen@dto.tno.nl; **Jeffrey** Paul, Cranfield University, Beds, UK, p.j.jeffrey@cranfield.ac.uk; **Jenni** Leo, Koordinationsstelle MGU, Basel, leo.jenni@unibas.ch; **Jenny** Katharina, Institute of Biotechnology, Zurich, jenny@biotech.biol.ethz.ch; **Jespersen** Per Homann, Roskilde University, Roskilde, Denmark, phj@teksam.ruc.dk; **Jörin** Ernst, Zürcher Hochschule Winterthur; **Joss** Simon, University of Westminster, London NW1 3SR, UK, josss@wmin.ac.uk; **Jungbluth** Niels, ETH Zurich, jungbluth@uns.umnw.ethz.ch; **Jungck** M., Swiss Office of Public Health, Bern, Matthias.

Jungck@bag.admin.ch; **Jurt** Luzia, Ethnologisches Seminar der Universität Zurich, lujurt@ethno.unizh.ch

Kägi Wolfram, BSS, Basel, Wolfram.Kaegi@bss-basel.ch; **Kaiser** Christine, Winterthur, kaiser@access.ch; **Kaiser** Tony, ALSTOM Power Technology Ltd, Baden-Daettwil; **Kamber** Rainer, Basel, rainer.kamber@unibas.ch; **Kämpf** Charlotte, University of Karlsruhe, Germany, charlotte.kaempf@bau-verm.uni-karlsruhe.de; **Kapila** Sunita, Eastern and Southern Africa Office, Nairobi, Kenya, SKapila@idrc.or.ke; **Kasanen** Pirkko, TTS Institute, Helsinki, Finland, pirkko.kasanen@tts.fi; **Kastenholz** Hans G., Center of Technology Assessement in Baden-Württemberg, Stuttgart, Germany, hans.kastenholz@ta-akademie.de; **Katell** Daniel, EPFL, Lausanne; **Katz** Sarah, AL-SAM-Anti-Drug Organisation, Ramat-Aviv, Tel Aviv, Israel; **Kauffman** Joanne, Massachusetts Institute of Technology, Cambridge, USA, jmkauffm@mit.edu; **Kaufmann** Stefan, Konkordat der Schweizerischen Krankenversicherer, Solothurn; **Kaufmann-Hayoz** Ruth, University of Bern, rkaufmann@ikaoe.unibe.ch; **Keeley** Stuart, Bowling Green State University, Bowling Green, USA, skeeley@bgnet.bgsu.edu; **Keitsch** Martina, Norwegian University of Science and Technology, Trondheim, Norway, martina.keitsch@indecol.ntnu.no; **Kesselring** Annemarie, Geschäftsstelle SBK, Bern, kesselring.sbk@bluewin.ch; **Kiefer** Bernd, Trägerverein Ökostromlabel Schweiz, Zurich, Kieferpartners@access.ch; **Kiiza** Johnson, Regional Engineers Office, Ministry of Works, Morogoro, Tanzania, iteco@raha.com; **Kinnas** John N., Athens, Greece, jkinnas@hotmail.com; **Kirsten-Krueger** Monika, Psychiatrische Universitätsklinik Zurich, kirsten0@bli.unizh.ch; **Kiteme** Boniface P., Nanyuki, Kenya, b.kiteme@africaonline.co.ke; **Klaus** Philipp, Hinteregg, klaus@smile.ch; **Kleiber** Charles, Staatssekretariat GWF, Bern, Dorothea.Brand@gwf.admin.ch; **Kleijnen** Jos, NHS Center for Review and Dissemination, York, UK; **Klein** Julie Thompson, Wayne State University, Ypsilanti Michigan, USA, JKlein4295@aol.com; **Klein** Michael, Institut für Neue Medien, Frankfurt/Main, Germany; **Klug Arter** Marianne, Inst. f. Umweltwissenschaften, Zurich, maklugar@uwinst.unizh.ch; **Knörzer** Andreas, Bank Sarasin+Cie, Basel, andreas.knoerzer@sarasin.ch; **Kocher** Gerhard, Zentralsekretariat SGGP, Muri; **Koch-Wulkan** Pedro W., BSV, Bern, pedro.koch@bsv.admin.ch; **Koenig** Ilse, Austrian Federal Ministry of Science and Transport, Vienna, Austria, ilse.koenig@bmwf.gv.at; **Kok** Marcel T.J., Global Change, Bilthoven, The Netherlands, Marcel.Kok@rivm.nl; **Kolb** Daniel, Metron AG, Brugg; **Köllner** Thomas, ETH, Zurich, koellner@uns.umnw.ethz.ch; **Korczak** Dieter, Club for Health, Weiler, Germany, GP-Forschungsgruppe@t-online.de; **Kostecki** Michel, Université de Neuchâtel, michel.kostecki@seco.unine.ch; **Kovacevic** Tea, Faculty of electrical engineering and computing, Zagreb, Croatia, tea.kovacevic@fer.hr; **Kowalski** Emil, GNW, c/o Nagra, Wettingen, Kowalski@gnw.ch; **Kreuzer** Konradin, Forum für verantwortbare Anwendung der Wissenschaft, Flüh, kreuzer@magnet.ch; **Krieg** Fritz, BMG Engineering AG, Schlieren, fritz.krieg@bmgeng.ch; **Kronauer** Brigitte, Hamburg, Germany; **Krott** Max, Georg-August-Universität Göttingen, Germany, mkrott@gwdg.de; **Krupp** Helmar, ISI, Weingarten, Germany, krupp@isi.fhg.de; **Kubasek** Nancy, Bowling Green State University, Bowling Green, Ohio, USA, nkubase@cba.bgsu.edu; **Kudadeniya** G.M.G., Kandy, Sri Lanka; **Kuebler** Markus, Teachers Training College, Spiez, mkuebler@bluemail.ch; **Kulikauskas** Paulius, Byfornyelsesselskabet Danmark, Copenhagen, Denmark, pauliusk@counsellor.com; **Kübler** Olav, ETH, Zurich; **Kulozik** Ulrich, Technische Universität München, Freising, Germany, ukulozik@kas.com; **Kundert** Sonja, Fachhochschule Solothurn und Nordwestschweiz, Solothurn, sonja.kundert@fhso.ch; **Küng** Valentin, Küng-Biotech + Umwelt, Bern, valentin.kueng@kueng-biotech.ch; **Kunz** Ulrich, BUWAL, Bern, ulrich.kunz@buwal.admin.ch; **Künzi** Erwin, Interdisciplinary Centre for General Ecology, Vienna, Austria, kuenzi@gpr.at; **Künzli** Christine, University of Bern, christine.kuenzli@ikaoe.unibe.ch; **Kurath** Monika, I. VW-HSG, St. Gallen, monika.kurath@unisg.ch; **Küry** Daniel, Life Science AG, Basel, kueryd@ubaclu.unibas.ch; **Kvarda** Werner, Univ. of Agricultural Sciences Vienna, Austria, freiraum@mail.boku.ac.at; **Kyrtsis** Alexandros-Andreas, University of Athens, Greece, kyrtsis@compulink.gr

Lanz Marco, Chefarzt Psychiatrie-Zentrum Hard, Zurich; **Larcher** Marie-Therese, Uitikon; **Lawrence** Roderick J., Université de Genève, lawrence@uni2a.unige.ch; **Lebert** Maud, Zurich, maudl@bluewin.ch; **Ledergerber** Elmar, Zurich; **Lefevre** Pierre, Institute of Tropical Medicine, Antwerp, Belgium, plefevre@itg.be; **Lehmann** Martin F., ERZ, Zurich, martin.lemann@erz.stzh.ch; **Lehmann** Peter, sanu, Biel, plehmann@sanu.ch; **Lehmann Pollheimer** Daniel, Swiss National Science Foundation, Bern, lehmann@sppe.ch; **Lehvo** Annamaija, Academy of Finland, Helsinki, Finland, annamaija.lehvo@aka.fi; **Lenz** Roman, Nuertingen University for Applied Sciences, Nuertingen, Germany, lenzr@fh-nuertingen.de; **Lepori** Carlo, Suspi-Idsia, Manno, lepori@idsia.ch; **Leroy** Pieter, Dept. of Environmental Policy Sciences, HK Nijmegen, The Netherlands, p.leroy@bw.kun.nl; **L'Homme** Serge, University of Quebec at Montreal, St-Leonard (Quebec), Canada, d241240@er.uqam.ca; **Limpert** Eckhard, Swiss Federal Institute of Technology, Lindau, eckhard.limpert@ipw.agrl.ethz.ch; **Loibl** Marie Céline, Austrian Institute for Applied Ecology, Vienna, Austria, oekoinstitut.plan@ecology.at; **Löw** Simon, ETH Hönggerberg, Zurich, Loew@erdw.ethz.ch; Luder Roland, Thun; **Ludi** Eva, Centre for Development and Environment, Institute of Geography, Bern, ludi@giub.unibe.ch; **Lukasczyk** Christian, sign language interpreter, Zurich, clukasczyk@access.ch; **Lukesch** Robert, ÖAR-Regionalberatung GmbH, Fehring, Austria, lukesch@eunet.at; **Lunca** Marilena, EX Utrecht, The Netherlands, river@river.tmfweb.nl;

Lüscher Thomas, Universitätsspital, Zurich; **Lutter** Christina, Austrian Federal Ministry of Science and Transport, Vienna, Austria, maria-christina.lutter@ bmwf.gv.at; **Luttropp** Conrad, KTH, Machine Design, Stockholm, Sweden, conrad@damek.kth.se; **Lys** Jon Andri, KFPE, Bern, kfpe@sanw.unibe.ch

Mahagedara Nandana Kumara, Provincal Council Office, Polgolla, Sri Lanka; **Maier** Simone, University of St. Gallen, smaier@idheap.unil.ch; **Marchini** Denise, Fachhochschule Solothurn Nordwestschweiz, Oensingen, denise.marchini@ fhso.ch; **Markard** Jochen, EAWAG, Kastanienbaum, jochen.markard@eawag.ch; **Marks** David H., Center for Environmental Initiatives, Cambridge, USA, dhmarks@mit.edu; **Marksthaler** Erich, Phoenix Contact AG, Tagelswangen, phoenixcontact@bluewin.ch; **Marti** Karin, topos Marti & Müller, Zurich, topos@ access.ch; **Martin** Claude, WWF International, Gland, cmartin@wwfnet.org; **Martinez** Sylvia, MCO Biodiversity/Integrated Project Biodiversity SPPE, Basel, sylvia.martinez@unibas.ch; **Maruping** Mpoeakae, National University of Lesotho, mp.ramollo@nul.ls; **Maselli** Daniel, DEZA, Bern, daniel.maselli@ deza.admin.ch; Mathieu Nicole, Nanterre, France; **Mathys** Renata, Applied University of Bern, Burgdorf, renata.mathys@hta-bu.bfh.ch; **Mauch** Corine, ETHZ, Zurich, mauch@fowi.ethz.ch; **Mebratu** Desta, UNECA/RCID, Addis Ababa, Ethiopia, dmebratu@hotmail.com; **Mechkat** Cyrus, IUED, Genève, cyrus. mechkat@iued.unige.ch; **Meier** Christine, InputUmwelt, Zurich, inputumwelt@ gmx.ch; **Meier** Jürg, Novartis International AG, Basel, juerg.meier@group.novartis.com; **Meier** Werner, Meier und Partner AG, Weinfelden, w.meier@meierpartner.ch; **Meier-Ploeger** Angelika, University of Applied Sciences Fulda, Germany, profmp@aol.com; **Meile** Eugen, Städtische Werke Winterthur, eugen.meile@ win.ch; **Mertens** Claudia, Zürcher Tierschutz, Zurich, cmertens@access.ch; **Meskens** Gaston, SCK.CEN, Mol, Belgium, gmeskens@sckcen.be; **Messerli** Bruno, Geographisches Institut, University of Bern, messerli@giub.unibe.ch; **Messerli** Paul, Geographisches Institut, University of Bern, mep@giub.unibe.ch; **Messerli** Peter, University of Bern, Centre for Development and Environment, Antananarivo, Madagascar, messerli@dts.mg; **Messerschmitt** Anja, Novartis, Basel, anja.messerschmitt@sn.novartis.com; **Mettier** Thomas, ETH, UNS, mettier@uns.umnw.ethz.ch; **Mey** Hansjürg, Kehrsatz, h.mey@bluewin.ch; **Meyer** Verena, Physik-Institut, Zurich, vmeyer@physik.unizh.ch; **Meylan** Jean-Pierre, Swiss Council of Universities of Applied Sciences, Bern, jmeylan@edk.unibe.ch; **Michalik** Georg, ETH UNS, Zurich, michalik@uns.umnw.ethz.ch; **Michelsen** Gerd, University of Lüneburg, Germany, michelsen@uni-lueneburg.de; **Mieg** Harald A., ETH Zurich, mieg@uns.umnw.ethz.ch; **Mirenowicz** Jacques, Institut pour la Communication et l'Analyse des Sciences et des Technologies, Fribourg, jacques.mirenowicz@icast.org; **Mittelstrass** Jürgen, University of Konstanz, Germany, juergen.mittelstrass@uni-konstanz.de; **Mogalle** Marc, University of St.

Gallen, Marc.Mogalle@unisg.ch; **Möhler** Hans, University and ETH Zurich, mohler@pharma.unizh.ch; **Mohr** Arthur, BUWAL, Bern, arthur.mohr@ buwal.admin.ch; **Moore** Patrick, Green Spirit, Vancouver, Canada, patrickmoore@ home.com; **Morgenthaler** Jean-Jaques, Bern, jjm@swissonline.ch; **Morolo** Tselane, National Research Foundation, Pretoira, South Africa, tmorolo@nrf.ac.za; **Moser** Karin S., University of Zurich, Social Psychology Unit, Zurich, kmoser@ sozpsy.unizh.ch; **Mouron** Patrik, Eidg. Forschungsanstalt, Wädenswil, patrik.mouron@faw.admin.ch; **Mueller** Alois, SKAT Swiss Centre for Development Cooperation in Technology and Management, St. Gallen, alois.mueller@skat.ch; **Müller** Heinz K., Swiss Federal Veterinary Office, Bern, Heinz.K.Mueller@bvet.admin.ch; **Müller** Pius, Marketing Service GmbH, Zurich, market@spectraweb.ch; **Müller** Regula, topos Marti & Müller, Zurich, topos@access.ch; **Müller** Thomas, Basler Zeitung, Basel; **Müller** Werner, Bird Life Zurich, werner.mueller@birdlife.ch; **Müller-Böker** Ulrike, Geographisches Institut, Zurich, boeker@geo.unizh.ch

Nägeli Sibylle, Zweckverband für Abfallverwertung Bezirk Horgen, kva.horgen@ bluewin.ch; **Nanayakkara** Eddie, Egodauya Na, Moratuwa, Sri Lanka; **Nentwich** Michael, Institute of Technology Assessment, Vienna, Austria, mnent@ oeaw.ac.at; **Neu** Urs, Pro Clim, Bern, neu@sanw.unibe.ch; **Neuhaus** Gabriele, Novartis, Basel, gabriele.neuhaus@seeds.novartis.com; **Nicolier** Felix, Novartis, Basel, felix.nicolier@group.novartis.com; **Niederer** Susanne, IBM, Zurich; **Niederhauser** Rolf, University of Basel, rolfniederhauser@compuserve.com; **Nielsen** Kurt Aagaard, Roskilde University, Roskilde, Denmark, aagaard@ ruc.dk; **Nösberger** J., ETH Zurich, josef.noesberger@ipw.agrl.ethz.ch; **Nowotny** Helga, Collegium Helveticum, Zurich

Oberle Bruno M., ETH Zurich; **Obrist** Hans-Ulrich, c/o Collegium Helveticum, Zurich, huo@compuserve.com; **Odermatt** André, Geographisches Institut, Zurich, odermatt@geo.unizh.ch; **Oegerli** Thomas, Professur für Soziologie, Zurich, oegerli@soz.gess.ethz.ch; **Oetliker** Sybille, Hebdo, Bern; **Oja** Ahto, SEI-Tallinn, University of Tartu, Institute of Geography, Tallinn, Estonia, ahto@ seit.ee; **Oktay** Ertan, Sahilyolu Orhantepe Mahallesi Yakamoz Sok, Dragos/ Istanbul, Turkey; **Ömer** Brigitte, Österreichisches Institut für nachhaltige Entwicklung, Vienna, Austria, oin@boku.ac.at; **Osterwalder** Walter, Basler & Hofmann Consulting Engineer, Zurich, wosterwalder@bhz.ch; **Oswald** Jenny, Swiss Federal Institute of Technology, Zurich, oswald@uns.umnw.ethz.ch; **Ouboter** Stefan, Centre for Soil Quality Management and Knowledge Transfer, Gouda, The Netherlands, skb@cur.nl

Pachlatko Christoph, Schweiz. Epilepsie-Klinik, Zurich; **Paula** Michael, Bundesministerium für Wissenschaft und Verkehr, Vienna, Austria, paula@bmwf.gv.at; **Pauli** Daniela, Forum Biodiversität Schweiz, Bern, daniela.pauli@sanw.unibe.ch; **Peltenburg**, Bern; **Penker** Marianne, Institute of Agricultural Economics, Vienna, Austria, penker@edv1.boku.ac.at; **Perera** H.U. Nishantha, Moratuwa, Sri Lanka; **Perincioli** Lorenz, Ingenieurbüro Energie+Umwelt, Goldiwil, l.perincioli@tcnet.ch; **Perren** Sonja, Institut Universitaire Kurt Bösch, Sion, sonja.perren@ikb.vsnet.ch; **Perritaz** N., BUWAL, Bern, nicolas.perritaz@buwal.admin.ch; **Petersen** Holger, University of Lüneburg, Germany, hpetersen@uni-lueneburg.de; **Pillet** Line, Swiss Science Agency, Bern, line.pillet-mevillot@gwf.admin.ch; **Ping** Xiao, Wuhan East Lake High Technology Group CO.,LTD, Wuhan City, Hubei Province, China, xiaoping16@hotmail.com; **Pivot** Agnès, NSS/CNRS, Nanterre, France, apivot@u-paris10.fr; **Plattner** Rolf M., Plattner Schulz Partner, Basel, pspag@pspag.ch; **Pohl** Christian, Collegium Helveticum, Zurich, pohl@collegium.ethz.ch; **Pok** Judith, Gynäkologie USZ, Zurich; **Pokorny** Doris, Biosphärenreservat Rhön, Oberelsbach, dpokorny@t-online.de; **Popow** Gabriel, LIB Strickhof, Lindau; **Portmann** Heidi, Gewaltfreie Aktion Kaiseraugst, Arlesheim, hportmann@datacomm.ch; **Pradhan** Leela, Tribhuvan University, Kathmandu, Nepal; **Protzen** Jean-Pierre, University of California, Berkeley, USA, protzen@socrates.berkeley.edu

Rabelt Vera, Umweltbundesamt, Berlin, Deutschland, vera.rabelt@uba.de; Ragaz Cheri, Zurich, chragaz@uwinst.unizh.ch; **Rais** Mohammad, National Institute of Science Technology and Development Studies, New Delhi, India, mohammad_rais@hotmail.com; **Raju** Kesiraju Venkata, Institute of Rural Management, Gujarant, India, kvr@fac.irm.ernet.in; **Ramsden** Jeremy, University of Basel, j.ramsden@unibas.ch; **Randegger** Johannes R., Novartis Services AG, Basel, bugldalania@bluewin.ch; **Rankine** Hitomi, National Institute of Higher Education, Trinidad and Tobago, niherst@opus.co.tt; **Rauber** Margit, Gwatt, rauber.gmbh@bluewin.ch; **Rauschmayer** Felix, München, Germany, rauschma@rz.uni-leipzig.de; **Rege** Colet Nicole, Université de Genève, Nicole.RegeColet@rectorat.unige.ch; **Regev** Gil, EPFL, Lausanne, gil.regev@epfl.ch; **Rehmann-Sutter** Christoph, Institut für Geschichte und Ethik der Medizin, Basel, christoph.rehmann-sutter@unibas.ch; **Reichl** Franz, Vienna University of Technology, Vienna, Austria, Franz.Reichl@tuwien.ac.at; **Reinhardt** Ernst, ecoprocess, Zurich, ernst.reinhardt@ecoprocess.ch; **Reiter** Wolfgang L., Bundesministerium f. Wissenschaft und Verkehr, Vienna, Austria, wolfgang.reiter@bmwf.gv.at; **Richardson** Phil, Geoscience Group, Leicestershire, UK, philR@quantisci.freeserve.co.uk; **Rieder** Peter, ETH Zentrum, Zurich, rieder@iaw.agrl.ethz.ch; **Rigon** Sandra, ORL-Institut, Zurich, rigon@orl.arch.ethz.ch; **Ringger** Heini, unicommunications, Zurich; **Rios** Estela, University of Quebec at Montreal, St-

Leonard, Quebec, Canada, rios.marta-estela@uqam.ca; **Ritz** Christoph, ProClim, Bern, ritz@sanw.unibe.ch; **Robert** Guy, Aix-en-Provence, France, grbrt@club-internet.fr; **Röllinghoff** Andreas, Research Manager of Method and Didactics of New Learning Technologies, Préverenges, rollinghoff@ndit.ch; **Roos** Andreas, Managed Care Winterthur Versicherung, Winterthur; **Roost Vischer** Lilo, Ethnologisches Seminar, Basel, lilo.roost-vischer@unibas.ch; **Rossel** Pierre, EPFL, Lausanne; **Rosselli** Walter, AR-FNP, Lausanne, wrossell@bluewin.ch; **Rössner** Petra, Kordinationsbüro Kulturlandschaftsforschung, Vienna, Austria, petra.roessner@klf.at; **Roth** Johann, Roth & Partner GmbH, Karlsruhe, Germany, irpka@swol.de; **Roux** Michel, Landwirtschaftliche Beratungszentrale Lindau, michel.roux@wsl.ch; **Ruckstuhl** Kilian, University of Zurich, k.ruckstuhl@access.unizh.ch; **Ruddy** Thomas, FH Solothurn NW-Schweiz, Olten, ruddyconsult@imailbox.com; **Rüegg** Walter, Eberhard Recycling AG, Kloten, walter.rueegg@eberhard.ch; **Rüegger** Heinz, Stiftung Diakoniewerk Neumünster, Zollikerberg; **Ruhnau** Eva, Humanwissenschaftliches Zentrum LMU München, München; **Rusterholz** Hans-Peter, Institut fur Natur-, Landschafts- und Umweltschutz der Uni Basel, hans-peter.rusterholz@unibas.ch; **Ryser** Walter, Rytec AG, Munsingen, rytec@swissonline.ch

Sabev Marinette, Science et Société, Petit-Lancy, sabev@cern.ch; **Saemann** Ralph, SATW, Basel; **Sailer** Michael, Öko-Institut e. V., Darmstadt, Germany, e.langenbach.rs@t-online.de; **Salembier** Pascal, GRIC-IRIT, Castanet-Tolosan, France, pgeslin@isp.fr; **Sarwar** Ivan, c/o Mosharaf Hossain, Dhaka, Bangladesh, ivansarwar@ieee.org; **Sauvain** Paul, BEREG, Bruson, serec.brus@gve.ch; **Scaroni** Fiorenzo, SUPSI-Direzione, Manno, scaroni@ti-edu.ch; **Schaffner** Beat, Geografisches Institut, University of Bern; **Schaltegger** Stefan, University of Lüneburg, Germany, schalteg@uni-lueneburg.de; **Schaub** Martin, CT Umwelttechnik AG, Winterthur, martin.schaub@ctu.ch; **Scheiwiller** Thomas, Pricewaterhouse Coopers, Zurich, thomas.scheiwiller@ch.pwcglobal.com; **Scheller** Andrea, Zurich, scheller@cepe.mavt.ethz.ch; **Schenk Wenger** Kaarina, BUWAL, Bern, kaarina.schenk@buwal.admin.ch; **Schenkel** Walter, Muri&Partner, Zurich, schenkel@pwi.unizh.ch; **Schenker** Franz, Geologische Beratung, Meggen, fsgeolog@tic.ch; **Schenler** Warren W., Swiss Federal Institute of Technology, Zurich, schenler@pst.iet.mavt.ethz.ch; **Scheringer** Martin, Swiss Federal Institute of Technology, Zurich, scheringer@tech.chem.ethz.ch; **Scheuermann** Michael, Albert-Ludwigs-Universität, Freiburg, Germany, scheuerm@psychologie.uni-freiburg.de; **Schibli** Daniela, EAWAG, Dubendorf, daniela.schibli@eawag.ch; **Schiess-Bühler** Corina, LBL, Lindau, corina.schiess@lbl.ch; **Schläpfer** Andreas, Swiss Re, Zurich, andreasL_schlaepfer@swissre.com; **Schlup** Michael, University of Basel, Röschenz, schlup00@stud.unibas.ch; **Schmid** C., University of Bern, schmid@giub.unibe.ch; **Schmid-Schönbein** Claudia, KS Graduate Business

School St. Gallen, class@e2mc.com; **Schmid-Schönbein** Oliver, E2 Management Consulting AG, Zurich, oliss@e2mc.com; **Schmidthaler** Franz, Healthcare Management Initiative, Fontainebleau, France; **Schmied** Barbara, ETH Zurich, D-UMNW, HCS, Zurich, barbara.schmied@umnw.ethz.ch; **Schmithüsen** Franz, Zurich, schmithuesen@fowi.ethz.ch; **Schneidewind** Uwe, K.Ossietzky Universität Oldenburg, Germany, uwe.schneidewind@uni-oldenburg.de; **Schnell** Klaus-Dieter, Institute for Public Services and Tourism, St. Gallen, klaus-dieter.schnell@unisg.ch; **Schnetzler** Rita, Zurich, ritschnetzler@dplanet.ch; **Scholz** Roland W., ETH, Swiss Federal Institute of Technology Zurich, scholz@uns.umnw.ethz.ch; **Schönherr** Hildegard, Berlin, Germany, Hildegard.Schoenherr@t-online.de; **Schönlaub** Hans P., Geological Survey of Austria, Vienna, Austria, hpschoenlaub@cc.geolba.ac.at; **Schönmann** Emil, KVL, Dietlikon; **Schorderet** Daniel F., CHUV, Lausanne, Daniel.Schorderet@chuv.hospvd.ch; **Schrader** Christoph, FACTS, Zurich; **Schrämli** Ruedi, Swiss Re, Zurich; **Schrefel** Christian, 17&4 Organisationsberatung GmbH, Vienna, Austria, christian.schrefel@17und4.at; **Schreier** Esther, MCO Biodiversity, Basel, schreier@uwinst.unizh.ch; **Schübel** Hubert R., Consultant for Organizational Psychology, Stuttgart, Germany, Schuebelhr@aol.com; **Schubinger** P. August, Paul Scherrer Institute, Villigen PSI, august.schubinger@psi.ch; **Schueler** Judith, University of Maastricht, The Netherlands, ja.schueler@barneveld.com; **Schüpbach** Erwin, quadra, Dussnang, winuschuepbach@bluewin.ch; **Schüpbach** Hans, LIB Strickhof, Lindau, hans.schuepbach@lbl.ch; **Schwarz** Astrid E., TU München, Freising, Astrid.Schwarz@spectraweb.ch; **Schweizer** Peter, MethoSys GmbH, Zurich, schweizer.peter@active.ch; **Seiler** Benno, econcept, Zurich, bs@econcept.ch; **Seiler** Hansjörg, Bundesgericht/Universität Freiburg, Münsingen, hansjoerg.seiler@gmx.ch; **Sell** Joachim, Swiss Federal Institute of Technology, Zurich, sell@uns.umnw.ethz.ch; **Semadeni** Marco, Swiss Federal Institute of Technology, Zurich, semadeni@uns.umnw.ethz.ch; **Sennheiser** Jörg, Sennheiser, Wedemark, Germany, sennjse@sennheiser.com; **Shmuel** Burmil, Faculty of Architecture and Town planning, Technion, Haifa, Israel, arshmuel@techunix.technion.ac.il; **Skorupinski** Barbara, Institute for Social Ethics, Zurich, baslc@access.unizh.ch; **Smoliner** Christian, Federal Ministery for Science and Transport, Vienna, Austria, christian.smoliner@bmwf.gv.at; **Smrekar** Otto, Redaktion Gaia, Basel; **Somermann** Américo, Escola do Futuro, University of São Paulo, Bresil, cetrans@futuro.usp.br; **Sommerfeld** Peter, Fachhochschule Solothurn und Nordwestschweiz, Solothurn, peter.sommerfeld@fhso.ch; **Sorg** Jean-Pierre, Swiss Federal Institute of Technology, Zurich, sorg@waho.ethz.ch; **Sotoudeh** Ariane, BUWAL, Bern, ariane.sotoudeh@buwal.admin.ch; **Spaapen** Jack B., sci_Quest, research agency for S&T Policy, SB Amsterdam, The Netherlands, jbspaa@xs4all.nl; **Spichiger** Ursula, ETHZ-Technopark, Zurich, uspi@chemsens.pharma.ethz.ch; **Spiess**, Migros Genossenschaftsbund MGB, Zurich, roman.spiess@

mgb.ch; **Spörri** Uli, Universitätsspital, Zurich; **Stadie** Marion, Dresden, Germany, MStadie@t-online.de; **Stalder** Andreas, BUWAL, Bern, andreas.stalder@buwal.admin.ch; **Stalder** Ueli, University of Bern, stalder@giub.unibe.ch; **Stamm** Christian, Swiss Federal Institute of Technology, Schlieren, stamm@ito.umnw.ethz.ch; **Stauffacher** Michael, Zurich, stauffacher@uns.umnw.ethz.ch; **Stauffacher** W., Basel, Werner.Stauffacher@unibas.ch; **Steffany** Frank, Universität zu Köln, Germany, steffany@meteo.Uni-Koeln.DE; **Steger**, Jean-Pierre, Berner Fachhochschule, Burgdorf, jean-pierre.steger@hta-bu.bfh.ch; **Steiner** Andreas, Wermatswil, ansteiner@access.ch; **Steiner** Regula, ETH-Zentrum, HCS, Zurich, steiner@uns.umnw.ethz.ch; **Steinmann** Walter, RBI Recyclingbetrieb für die Industrie AG, Bazenheid, rbirecycling@bluewin.ch; **Steurer** J., Universitätsspital, Zurich, johann.steurer@dim.usz.ch; **Stoll** Susanne, ETH Zurich, Dubendorf, susanne.stoll@eawag.ch; **Stoltenberg** Ute, University of Lüneburg, Germany, stoltenberg@uni-lueneburg.de; **Storch** Maja, Pädagogisches Institut Universität Zurich, storch@paed.unizh.ch; **Strebel** Urs, Kreisspital Männedorf; **Stricker** Hermann W., Pfizer AG /Prasident VIPS, Zurich; **Stricker** Thomas, Basel, thomas.stricker@pentapharm.com; **Stucki** Gerhard, Ciba Spezialitätenchemie AG, Pratteln, gerhard.stucki@cibasc.com; **Stuecheli** Alexander, ZHW Zurich University of Applied Sciences Winterthur, stu@zhwin; **Stuhler** Elmar A., Technische Universität Munchen, Germany, stuhler@pollux.weihenstephan.de; **Stünzi** Jürg, Stadt Winterthur, stuenzi.goe@bluewin.ch; **Suhr** Nelson Julie, University of Utah, Salt Lake City, USA, jnelson1@slkc.uswest.net; **Sundgren** Jan-Eric, Chalmers University, Göteborg, Sweden, jan-eric.sundgren@adm.chalmers.se; **Sundin** Nils-Göran, Collegium Europaeum, Stockholm, Sweden, nils-goran.sundin@mailcity.com; **Suter** Hans, Gekal/KVA Buchs, Buchs, h.suter@kva-buchs.ch; **Suter** Jürg, AWEL, Zurich; **Suter** Karin, AGS/ETH Zurich, karin.suter@sl.ethz.ch; **Suter** Roger, Alliance for Global Sustainability, Zurich; **Syfrig** Josef, B.I.C.S./BioTeCH forum, Basel, josef.syfrig@unibas.ch

Tanaka Keiichi, Nihon University, Tokyo, Japan, harasawa@nies.go.jp; **Tanner** Carmen, University of Freiburg, carmen.tanner@unifr.ch; **Tanner** Isabel, ETH Zurich, i_tanner@yahoo.com; **Tanner** Marcel, Swiss Tropical Institute, Basel, tanner@ubaclu.unibas.ch; **Taroni** Franco, Université de Lausanne, Franco.Taroni@inst.hospvd.ch; **Telli** Sahure Gonca, Maltepe University, Istanbul, Turkey, goncatelli@superonline.com; **Terwiesch** Peter, Baden-Daettwil; **Tewari** Anil Kumar, Nesco, Allahabad, India; **Thierstein** Alain, IDT-HSG, University of St. Gallen, alain.thierstein@unisg.ch; **Thierstein** Hans R., Geologisches Institut ETH Zentrum, Zurich, thierstein@erdw.ethz.ch; **Tolunay** Renate, Vienna University of Economics and Business Administration, Vienna, Austria, tolunay@wu-wien.ac.at; **Tretter** Felix, München, Germany; **Trezzini** Giampiero, Lausanne, giampiero.trezzini@urbanet.ch; **Truffer** Bernhard, Swiss Federal Institute for

Environmental Science and Technology, Kastanienbaum, truffer@eawag.ch; **Tsujigado** Makoto, Kogakuin University, Tokyo, Japan, tsuji@ee.kogakuin.ac.jp; **Tuinstra** Willemijn, Wageningen Agricultural University, Wageningen, The Netherlands, willemijn.tuinstra@wimek.cmkw.wau.nl; **Tytarenko** Lydia, The Voice of Ukraine, Cherkassy, Ukraine

Ulbrich Susan, Zurich, ulbrich@uns.umnw.ethz.ch; **Ulli-Beer** Silvia, University of Bern, ulli@ikaoe.unibe.ch; **Ulrich** Markus, UCS Ulrich Creative Simulations, Zurich, ucs@access.ch; **Umbricht** Michael, Professur Natur- und Landschaftsschutz, Uznach, umbricht@nls.umnw.ethz.ch

Vahtar Marta, Institute for Integral Development and Environment, Domzale, Slovenia, marta.vahtar@guest.arnes.si; **Valsangiacomo** Antonio, University of Bern, valsan@sis.unibe.ch; **van de Kerkhof** Marleen, Institute for Environmental Studies, Amsterdam, The Netherlands, marleen.van.de.kerkhof@ivm.vu.nl; **van der Merwe** Louis, Centre for Innovative Leadership, The Hague, The Netherlands, louis@cil.net; **van Emburg** Peter, EV Leiden, The Netherlands, p_r_van_emburg@wxs.nl; **van Hoorn** Thessa, TNO Institute of Strategy, JA Delft, The Netherlands, vanHoorn@stb.tno.nl; **van Veen** H. Johan; Varlez Sylvie, Federal Planning Bureau, Bruxelles, Belgium, sv@plan.be; **Velosa da Silva** Claudia, Deutscher Sparkassen- u. Giroverband, Bonn, Germany, Claudia.Velosadasilva@DSGV.de; **Vermeulen** W.J.V., Utrecht University, Utrecht, The Netherlands; **Vetter** Willhelm, Medizinische Poliklinik, Zurich; **Veya** Elisabeth, Science et Cité, Bern; **Villa** Alessandro, Université de Lausanne, Alessandro.Villa@iphysiol.unil.ch; **Vintges** M-G. M., The Hague, The Netherlands, vintges@nwo.nl; **Voegeli** Hans F., Zürcher Kantonalbank, Zurich, hans.voegeli@zkb.ch; **Vogel** René, PricewaterhouseCoopers, Bern, rene.vogel@ch.pwcglobal.com; **von Gruenewaldt** Gerhard, National Research Foundation, Pretoria, South Africa, gerhard@nrf.ac.za; **von Lutterotti** Nicola, Frankfurter Allgemeine Zeitung, Celigny, nicolutter@cs.com; **von Reding** Walter, MBA Network Economics, Steinen, waltervonreding@csi.com; **von Waldkirch** Thomas, Technopark, Zurich, waldkirch@technopark.ch; **Vonlaufen** Adrian, ENTSO TECH AG, Winterthur, etag@entsotech.ch

Wacker Corinne, University of Zurich, wacker@ethno.unizh.ch; **Wackers** G.L., University of Maastricht, The Netherlands, g.wackers@tss.unimaas.nl; **Waeber** Roger, Swiss Federal Office of Public Health, Bern, roger.waeber@bag.admin.ch; **Wäger** Patrick, Swiss Federal Laboratories for Materials Testing and Research, St. Gallen, patrick.waeger@empa.ch; **Wagner** Conrad, WestStart-CALSTART, Alameda, USA, w@gner.ch; **Wagner** Werner, Novartis/Valorec AG, Basel, werner.wagner@sn.novartis.com; **Waldner** Rosmarie, Zurich, 100606.756@

compuserve.com; **Waldvogel** Albert, Vizepräsident Forschung ETH, Zurich, albert.waldvogel@sl.ethz.ch; **Waldvogel** Francis, Conseil des EPF, Zurich; **Walin** Laura, Research Programme on Biological Functions/Institute of Biotechnology, University of Helsinki, Finland, laura.walin@helsinki.fi; **Walker** Beat, Deponie Teuftal AG, Frauenkappelen, deponie@teuftal.ch; **Walser** Manfred, Institute for Public Services and Tourism, St. Gallen, manfred.walser@unisg.ch; **Walter** Alexander, ETH Zurich, waltera@student.ethz.ch; **Walter** Thomas, Eidgenössische Forschungsanstalt für Agrarökologie und Landbau, Zurich, thomas.walter@fal.admin.ch; **Walther** Jessica, Marketing Service Pius Müller GmbH, Zurich, jrw@smile.ch; **Wamelink** Frank, sci_Quest, Maarssen, The Netherlands, wamelink@chem.uva.nl; **Wasem** Karin, ETH Zurich, UNS, Zurich, wasem@uns.umnw.ethz.ch; **Weber** Lukas, Swiss Federal Institute of Technology, Zurich, lukas.weber@cepe.mavt.ethz.ch; **Weber** Matthias, JRC-IPTS, Sevilla, Spain, matthias.weber@jrc.es; **Weber** Olaf, Swiss Federal Institute of Technology, Zurich, weber@uns.umnw.ethz.ch; **Wehrli** Bernhard, EAWAG, Kastanienbaum, bernhard.wehrli@eawag.ch; **Wehrli** Christoph, NZZ, Zurich; **Wehrli-Schindler** Brigit, Zurich; **Weidner** Helmut, Wissenschaftszentrum Berlin für Sozialforschung, Berlin, kollande@medea.wz-berlin.de; **Weiss** Martina, Collegium Helvetivum, Zurich, weiss@collegium.ethz.ch; **Welti** Myrta, Science et Cité, Bern, info@science-et-cite.ch; **Werner** Frank, EMPA, Dübendorf, frank.werner@empa.ch; **Wespe** Rolf, BUWAL, Bern; **Wickramasinghe** S.M., Moratuwa, Sri Lanka; **Widmer** Hans, Zürcher Kantonalbank, Zurich, hans.widmer@zkb.ch; **Wiesmann** Urs, University of Bern, wiesmann@giub.unibe.ch; **Wigum** Kristin Stoeren, Trondheim, Norway, kriswig@design.ntnu.no; **Wijesundera** W.M.S, Kandy, Sri Lanka; **Wilhelm** Beate, University of St. Gallen, beate.wilhelm@unisg.ch; **Willemsen** Ariane, BUWAL, Bern, ariane.willemsen@buwal.admin.ch; **Willnauer** Susanne, Migros Genossenschaftsbund MGB, Zurich, Susanne.Willnauer@mgb.ch; **Wils** Jean-Pierre, University of Nijmegen, The Netherlands, J-P.Wils@theo.kun.nl; **Wokaun** Alexander, Paul Scherrer Institute, Villigen PSI, alexander.wokaun@psi.ch; **Wolf** Markus, University of Zurich; **Wolfensberger** Ruth, Ethnologisches Seminar, Basel, lilo.roost-vischer@unibas.ch; **Woschnack** Ute, Gebenstorf, woschnack@uns.umnw.ethz.ch; **Wullkopf** Uwe, Institut Wohnen und Umwelt GmbH, Darmstadt, Germany, u.wullkopf@iwu.de; **Wust** Sebastian, EPFL-DA-IREC, Lausanne, sebastien.wust@epfl.ch; **Wüstenhagen** Rolf, IWÖ-HSG, St. Gallen, rolf.wuestenhagen@unisg.ch; **Wüthrich** Christoph, University of Basel, christoph.wuethrich@unibas.ch; **Wüthrich** Peter, Bundesamt für Sport Magglingen, peter.wuethrich@baspo.admin.ch; **Wyder** Rita, BUWAL, Bern; **Wyss** Kaspar, Swiss Tropical Institute, Basel, wyssk@ubaclu.unibas.ch; **Wyss** Otto, Dir. Sportbahnen Hasliberg-Käserstatt, Hasliberg Twing

Yémadji N'Diekhor, Centre de Support en Santé International, N'Djaména, Chad, cssiitsn@intnet.td; **Yetergil** Devrim, ETH Zurich, yetergil@umnw.eth.ch; **Yonkeu** Samuel, Inter-State School of Engineers of Rural Equipment, Ouagadougou 03, Burkina Faso, syonkeu@yahoo.fr

Zahner Adrian, Dialog Ethik, Zurich; **Zangger** Annika, ETH Zurich, brunnand@student.ethz.ch; **Zeh** Walter, BUWAL; Swiss Ag. for Envir., Forest and Landscape, Bern, walter.zeh@buwal.admin.ch; **Zehnder** Stefan, Marketing Service Pius Müller GmbH, Zurich, stefan.zehnder@gmx.ch; **Zimmermann** Jennifer, WWF Schweiz, Zurich, Jennifer.Zimmermann@wwf.ch; **Zoglauer** Michael, TIWAG, Innsbruck, Austria, michael.zoglauer@tiwag.at; **Zucker** Betty, Gottlieb Duttweiler Institut, Rüschlikon, betty.zucker@gdi.ch; **Zuidema** Piet, Nagra, Wettingen, zuidema@nagra.ch; **Zweibrücken** Klaus, Hochschule Rapperswil, kzweibru@hsr.ch

Appendix D

List of Contributors

Italized references in this book are to the pre-conference Workbook texts:

Transdisciplinarity: Joint Problem-Solving among Science, Technology and Society. Proceedings of the International Transdisciplinarity 2000 Conference (Zurich: Haffmans Verlag, 2000).

- Workbook I: Dialogue Sessions and Idea Market, ed. R. Häberli, et al.
- Workbook II: Mutual Learning Sessions, ed. R. Scholz, et al.

They are available at <http://www.transdisciplinarity.ch>

Aagaard Nielsen Kurt et al., *Democratic Challenges in the Risk Society, D06*, Hus 11.2 Ruc, P.Box 260, 4000 Roskilde, Denmark, pa@teksam.ruc.dk, I 224–229.

Abbasi S.A. et al., *Transdisciplinary Research in India – Why it is Rarely Done and Why it even more Rarely Suceeds: A Likely third World Scenario, D06*, Pondicherry University, Centre for Pollution Control, Kalpet, Pondicherry, India, prof_abbasi@vsnl.com, I 263–267.

Abou-Kahled Omar et al., *New Concepts for Continuing Education – Challenge to the Entire Education System, M07*, Ecole d'Ingénieurs de Fribourg, Boulevard de Pérolles 80, 1705 Fribourg, aboukhal@di.epfl.ch, II 130–134.

Adey Bryan et al., *Societal Benefits of Preservation Strategies for Civil Infrastructure, I02*, Swiss Federal Institute of Technology, Institut de Statique et Structures, MCS-EPFL GC B2, 1015 Lausanne, Bryan.Adey@epfl.ch, I 620–624.

Adi Alpheus Bongo C., *Globally Sustainable Development, I02*, Restral Consulting Limited, c/o Mr. Kenneth Amaeshi, Plot 1664, Oyin Jolayemi Street, Victoria Island Lagos, Nigeria, nesg@hyperia.com, I 585–589.

Affolter Stefanie et al., *Stories of Sustainability. Success and Failure in Sustainability Processes, M12*, Kant. Amt für Umweltschutz, Klosterstrasse 31, 6002 Luzern, gallati@die-luft.ch, II 253–257.

Aiking Harry et al., *PROFETAS: Protein Foods, Environment, Technology and Society, D08*, Vrije Universiteit, Institute for Environmental Studies, De Boelelaan 1115, Amsterdam, Netherlands, harry.aiking@ivm.vu.nl, I 326–330.

Aliò Maria Àngels et al., *In Search of New Knowledge for Urban Areas: The Potential of Citizen Participation in the Implementation of the Local Agenda 21 in Municipalities of the Barcelona Metropolitan Area, M12*, Universitat de Barcelona, Departament de Geografia Humana, C/ Baldiri Reixac, s/n., O8028 Barcelona, Spain, angelsa@trivium.gh.ub.es, II 246–249.

Amhar Fahmi, *Innovation Infrastructure as Transdisciplinarity Problem, D07*, Center for International Trends and Alternative Strategies Studies, Bakosurtanal, Jl. Jakarta-Bogor Km. 46, Cibinong, Indonesia, famhar@hotmail.com, I 280–281.

Baechler Günther, *Water Management and Conflict. Transformation in the Blue Nile Basin. An action-oriented Stakeholder Approach, D12*, Swiss Peace Foundation, Institute for Conflict Resolution, PO, 3000 Bern, baechler@swisspeace.unibe.ch, I 482–484.

Bährer Sabine et al., *The therapeutic interventionprogram for demented persons, their partners and closed relatives, I02*, Psychiatric University Hospital, Wilhelm Klein-Strasse 27, 4025 Basel, sabine.baehrer@datacomm.ch, I 648–651.

Balyeku Andrew, *Jinja Case Study, M12*, Jinja Municipal Council, P.O Box 45, Jinja, Uganda, jmc@swiftuganda.com, II 250–252.

Bassand Michel et al., *Metropolization, Ecological Crisis and Sustainable Development: Water Management and Disadvantaged Neighborhoods in Ho Chi Minh-City, Vietnam, M11*, Swiss Federal Institute of Technology/IREC, Av. Eglise Anglaise 14, PO 555, 1001 Lausanne, bassand@irec.da.epfl.ch, II 206–210.

Bassand Michel et al., *Society and its Actors – A Transdisciplinary Approach, D02*, Swiss Federal Institute of Technology/IREC, Av. de l'Eglise Anglaise 14, PO 555, 1001 Lausanne, bassand@irec.da.epfl.ch, I 64–67.

Baumann Max, *Intra-, Inter- and Transdisciplinarity in Health Care – the Lawyer's View, M06*, University of Zurich/Interdisziplinäres Institut für Ethik im Gesundheitswesen, Gloriastrasse 18, 8028 Zurich, info@dialog-ethik.ch, II 111– 113.

Baumann-Hölzle Ruth, *Health costs and benefits. Who costs and benefits in health care? M06*, Säntisstrasse 1, 8633 Wolfhausen, II 106–110.

Baumgartner Ruedi, *Mutual Learning in Intercultural Field Research: Sharing Research Findings with Rural Communities, M18*, ETH Zurich / NADEL, ETH Zentrum, 8092 Zurich, baumgartner@nadel.ethz.ch, addendum available directly from Baumgartner, by e-mail.

Baumgartner Thomas, *Mobility and Communication Videoconferencing or Face-to-Face Meetings? The Issue of Noise Control, D05*, Swiss Federal Institute of Technology, Natural & Social Science Interface, ETH Zentrum HAD, Haldenbachstrasse 44, 8092 Zurich, baumgartner@uns.umnw.ethz.ch, I 184–188.

Baur Bruno, *Effects of Recreational Activities on Ground Vegetation, Shrubs and Ground-Dwelling Invertebrates in a Suburban Forest, D03*, University of Basel,

Department of Integrative Biology, Section of Conservation Biology (NLU), St. Johanns-Vorstadt 10, 4056 Basel, baur@ubaclu.unibas.ch, I 87.

Bearth Thomas, *Language, Communication and Sustainable Development: A Neglected Area of Interdisciplinary Research and Practice, D07*, University of Zurich, Dept. of General Linguistics, Plattenstrasse 54, 8032 Zurich, Thomas_Bearth@compuserve.com, I 170–175.

Becker Egon, *Sustainability – A Cross-Disciplinary Concept for Social Transformations, D01*, J.W.-Goethe University, Dept. of Educational Sciences, Senckenberganlage 31, 60325 Frankfurt am Main, Germany, e.becker@em.uni-frankfurt.de, I 29–31.

Bernhard Hans-Peter, *Contribution of Genetic Engineering to Sustainable Agriculture? M02*, Novartis Services AG, Basel Works K-25.1.02, 4002 Basel, hans-peter.bernhard@sn.Novartis.com, II 46.

Bierens de Haan Camille et al., *Oxymoron, a Non-Distance Knowledge Sharing Tool for Social Science Students and Researchers, D05*, Institut universitaire Kurt Bösch, Case Postale 4176, 1950 Sion, camille.bdehaan@ikb.vsnet.ch, I 191–195.

Bill Alain, *China Energy Technology Program (CETP), M01*, ALSTOM Power Technology Ltd (formerly at ABB Corporate Research Ltd), 5405 Baden-Dättwil, alain.bill@power.alstom.com, II 20–21.

Bocharnikov Vladimir N., *GIS allows Russian Indigenous Sustainable Tourism, M05*, Pacific Institute of Geography, 690041 Vladivostok, Russia, sergeikr@online.vladivostok.ru, II 85–88.

Bonazzi Achille et al., *Complexity and Transdisciplinarity for Environmental Education, M07*, University of Parma, Italian Center for Environmental Education, Via Cavestro 14, 43100 Parma, Italy, cirea@ipruniv.cce.unipr.it, II 135–137.

Böttcher Sabine et al., *Successful Ecological Innovations through Strategic Technology Management, M09*, University of St. Gallen, Institute for Technology Management, Unterstr. 16, 9000 St. Gallen, sabine.boettcher@unisg.ch, II 160–165.

Brenner Joseph E., *The Psychology of Transdisciplinarity, D02*, OEUF, PO BOX 235, 1865 Les Diablerets, jebrenner@compuserve.com, I 72–77.

Browne Neil et al., *The Rigidity of An Attorney's Personality: Can Legal Ethics Be Acquired? I02*, Bowling Green State University, Bowling Green, USA, nbrown2@cba.bgsu.edu, I 640–644.

Buchecker Matthias, *Ways to Encourage Local Residents to Participate in Landscape Development, M13*, Swiss Federal Institute for Forest, Snow and Landscape Research, Birmensdorf, matthias.buchecker@wsl.ch, II 289–292.

Büchel Dominik, *Dialogue on Genetic Testing: A Successfull Participatory Experiment, M17*, Locher, Brauchbar & Partner, Wettsteinallee 7, 4058 Basel, Buechel@lbp.ch, II 318–320.

Buchinger Eva, *Transdisciplinarity: Experiences in Collaborations between Science and Citizens and Science and Policy, M17*, Austrian Research Centers Seibersdorf (ARCS), 2444 Seibersdorf, Austria, eva.buchinger@arcs.ac.at, II 340–343.

Bulych Yaroslav et al., *Science and Agro- and Ecotourism Development in Ukrainian Carpathians, M05*, State Dept. of Environmental Safety in Lviv Region, Stryiska 98, 290026 Lviv, Ukraine, root@envir.lviv.ua, II 93–97.

Burger Paul et al., *Transdisciplinary Training and Research: Experience from the Transdisciplinary Program of the Foundation "Man-Society-Environment" (MGU), D03*, University of Basel, Koordinationsstelle MGU, PO, 4002 Basel, burger@ubaclu.unibas.ch, I 84–86.

Burger Paul, *What Kind of Knowledge do we Gain in Inter- and Transdisciplinary Research and how do we Justify it? D03*, University of Basel, Koordinationsstelle MGU, PO, 4002 Basel, burger@ubaclu.unibas.ch, I 104–108.

Bütschi Danielle, *Mutual Learning Session on Participation: Towards Integration of Lay Expertise in the Debate on Technological and Scientific Developments, M17*, Conseil Suisse de la Science, Programme Technology Assessment, Inselgasse 1, 3003 Bern, danielle.buetschi@swr.admin.ch, II 312–314.

Bütschi Danielle, *The Integration of Lay Expertise in Technology Assessment: Swiss PubliForums as an Example, M17*, Conseil Suisse de la Science, Programme Technology Assessment, Inselgasse 1, 3003 Bern, danielle.buetschi@swr.admin.ch, II 326–330.

Caetano João Carlos et al., *On Transdisciplinarity in Organizational, Technological Change and Law, I01*, Universidade Aberta, Departamento de Organização e Gestão de Empresas, Rua Fernão Lopes 9, 1° Esq., 1000 Lisboa, Portugal, joaocaetano@mail.telepac.pt, I 528–533.

Canali Susan, *Visual/Experiential Learning Tools for Revealing a Holistic view of a Situation under Study, I01*, International Institut for Industrial Environmental Economics, Dag Hammarskjolds vag 3G, 224 64 Lund, Sweden, scanali@yahoo.com, I 556–558.

Carter Claudia et al., *Encouraging Interdisciplinary Research and Debate: Experience from the Concerted Action on Environmental Valuation in Europe (EVE) (June 1998–November 2000), D05*, University of Cambridge, Cambridge Research for the Environment, 19 Silver Street, CB3 9EP Cambridge, UK, cec29@cam.ac.uk, I 166–169.

Charlesworth Mark, *Sustainable Development: Transdisciplinary Research Programmes – The Most Important Transdisciplinary Research Question? D08*, Keele University, School of Politics, ST5 5BG Keele, UK, m.e.charlesworth@pol.keele.ac.uk, I 331–335.

Craye Matthieu et al., *Transdisciplinarity in Environment & Health Risk Assessment, I01*, University of Antwerpen, Research Centre on Technology, Energy and

Environment, Kleine Kauwenberg 12, 2000 Antwerpen, Belgium, matthieu.craye@ufsia.ac.be, be, I 544–549.

Cslovjecsek Markus, *Communication and Sign Systems – An Instruction Attempt for Transdisciplinary Communication, I01*, Gruengli 45, 4523 Niederwil, m.cs@bigfoot.com, I 495–500.

Cslovjecsek Markus, *Sound leaded Learning, M07*, Gruengli 45, 4523 Niederwil, m.cs@bigfoot.com, II 149–153.

Dahinden Urs et al., *Biotechnology: Dimensions of Public Concern in Europe, M02*, University of Zurich, Institut für Publizistikwissenschaft und Medienforschung, PO 507, 8035 Zurich, dahind@ipmz.unizh.ch, II 47–51.

De Mello Maria F. et al., *Transdisciplinary Evolution in Education: Contributing to the Sustainable Development of Society and of the Human Being, M19*, University of São Paulo, Escola de Futuro, Av. Prof. Lúcio Martins Rodrigues, Esq. C Cidade Universitária, 05508-900 São Paulo, Brasil, cetrans@futuro.usp.br, II 380–383.

Decker Michael, *The Origin of Transdisciplinarity. A proposal to organise an interdisciplinary expert group, D05*, Europäische Akademie zur Erforschung von Folgen wissenschaftlich-technischer Entwicklungen, Bad Neuenahr-Ahrweiler GmbH, PO 1460, D 53459 Neuenahr-Ahrweiler, Germany, michael.decker@dlr.de, I 206–210.

Defila Rico et al., *How can inter- and transdisciplinary Cooperation best be designed? Results from a comparative Survey of four Research Programmes, M20*, University of Bern, Interdisciplinary Centre for General Ecology, Falkenplatz 16, 3012 Bern, rico.defila@ikaoe.unibe.ch, II 391–395.

Demyanenko Valèry et al., *Methodological Approaches to the Stage Formation of Qualities of an Ecologically Educated Personality, D05*, Shevchenko 460, 18006 Cherkassy, Ukraine, is2000@gate.chiti.uch.net, I 179–183.

Di Giulio Antonietta et al., *An Assessment Instrument for inter- and transdisciplinary Research, D03*, University of Bern, Interdisciplinary Centre for General Ecology, Falkenplatz 16, 3012 Bern, digiulio@ikaoe.unibe.ch, I 109–110.

Di Giulio Antonietta et al., *Animating Transdisciplinarity – The Management of an integrated Project, I01*, University of Bern, Interdisciplinary Centre for General Ecology, Falkenplatz 16, 3012 Bern, digiulio@ikaoe.unibe.ch, addendum available directly from Di Giulio by e-mail.

Drilling Matthias, *The Delphi Technique: A Tool for Interdisciplinary and Transdisciplinary Research. Conclusions From a University Training Project, D06*, Hohere Fachschule im Sozialbereich, Abt. Weiterbildung, Dienstleistung, Forschung, Thiersteinerallee 57, 4053 Basel, mdrilling@hfsbb.unibas.ch, I 254–258.

Dürrenberger Gregor et al., *Focus Groups for Participatory Assessment Technique, M17*, Swiss Federal Institute for Environmental Science and Technology,

Division of Resource and Waste Management, Ueberlandstr. 133, 8600 Dübendorf, Gregor@eawag.ch, II 321–325.

Egger Gregory et al., *Conscious Development of the Krappfeld Region – First Steps Towards Transdisciplinarity, D08*, E.C.O. Institute for Ecology, Jungmeier KG, Burggasse 10, 9020 Klagenfurt, Austria, eco@aon.at, I 301–303.

Ehmayer Cornelia, *Cultural Landscapes and AGENDA 21 (CULT:AG), M12*, 17&4, Mariahilferstrasse 89/29, 1060 Vienna, Austria, cornelia.ehmayer@magnet.at, II 243–245.

Flüeler Thomas, *Mutual Learning on Radioactive Waste Management: Transcontinental Gridlock-Transdisciplinary Solution? M15*, Umweltrecherchen & -gutachten, Münzentalstr. 3, 5212 Hausen/AG, flueeler_urg@bluewin.ch, II 304–307.

Flury Manuel, *North-South Dialogue on Environmental Management «Learning from the Others», I02*, University of Bern, Interdisciplinary Centre for General Ecology, Falkenplatz 16, 3012 Bern, flury@ikaoe.unibe.ch, I 615–616.

Flury Manuel, *The Discussion Forum North-South reviewed – Lessons for effective exchange and transfer, M18*, University of Bern, Interdisciplinary Centre for General Ecology, Falkenplatz 16, 3012 Bern, flury@ikaoe.unibe.ch, II 358–362.

Förster Ruth et al., *The SAGUF-Network for Transdisciplinary Research, I02*, University of Basel, Co-ordination Centre Men-Society-Environment, Socinstr. 59 / Post Box, 4002 Basel, foersterruth@ubaclu.unibas.ch, I 645–647.

Förster Ruth, *Criteria for Training in Transdisciplinary Practice: Experiences from the MGU-Education Programm, D03*, University of Basel, Co-ordination Centre Men-Society-Environment, Socinstr. 59/Post Box, 4002 Basel, foerster ruth@ubaclu.unibas.ch, I 93–97.

Frischknecht Peter et al., *Teaching Transdisciplinarity in a First Semester Course, I02*, Swiss Federal Institute of Technology, Dep. of Environmental Sciences, 8092 Zurich, frischknecht@umnw.ethz.ch, I 627–631.

Fry Patricia E. et al., *Comparing Farmer's and Scientist's Views on Biodiversity and Soil Quality, D11*, Swiss Federal Institute of Technology, Institute for Terrestrial Ecology/Collegium Helveticum, Volkmarstr. 9, 8006 Zurich, fry@wawona.ethz.ch, I 411–415.

Füssler Jürg, *Flexibility mechanisms for greenhouse gas reduction (JI/CDM), An example of research and project implementation in a transdisciplinary framework, M01*, Ernst Basler + Partner AG, Mühlebachstrasse 11, 8032 Zurich, juerg.fuessler@ebp.ch, II 28–30.

Gandolfi Alberto F., *For a Better Management of Complex Systems: The European Academy for Politicans (EUCAP), I02*, Via Leoncavallo, 6614 Brissago, alberto.gandolfi@ch.pwcglobal.com, I 578–581.

Genske Dieter D. et al., *Water-talk: a transdisciplinary contribution to solve the water crisis, D12*, Swiss Federal Institute of Technology, Institut de Génie de l'environnement, 1015 Lausanne, dieter.genske@epfl.ch, I 477–481.

Gerber Alexander, *Action Research as an Instrument for Transdisciplinary Research in the Pilot Project "Cultural Landscape Hohenlohe", D11*, Projektgruppe Kulturlandschaft Hohenlohe, Universität Hohenheim, 70593 Stuttgart, Germany, gerberal@uni-hohenheim.de, I 406–410.

Gerber Mariette et al., *Nutrition and Disease Prevention: a Field for Transdisciplinarity, M04*, Centre de Recherche en Cancérologie INSERM-CRLC, Groupe d'Epidémiologie Métabolique, 34298 Montpellier, France, marietger@valdorel.fnclcc.fr, II 78–79.

Gerbilsky Lew, *The Transdisciplinary Nature of National Health and Environment Action Plans, M06*, c/o Dr. O.Kharytonov, Groenhoffweg 8, 24159 Kiel, Germany, Lew_Gerbilsky@excite.com, II 119–123.

Gersbach Klaus, *Traditional Fruit Trees in the Countryside: An Urgent Problem to be Solved, M13*, Agricultural Extension Center, Eschikon, 8315 Lindau, II 286–288.

Geslin Philippe et al., *Antropology and Ergonomics in Designing Innovations: Theoretical and Methodological Foundations of a Transdisciplinary Research, I01*, IMRA-SAD, B.P. 27, 31326 Castanet-Tolosan, France, pgeslin@toulouse.inra.fr, I 488–491.

Gfeller W. et al., *Radon: The Deadly Visitor, M10*, Swiss Office of Public Health, 3003 Bern, matthias.jungck@bag.admin.ch, II 183–186.

Ghosh Santosh, *Our Cities, Their Cities: Sustainable Development, M11*, Centre for built Environment, 2/5 Sarat Bose Road, 700020 Calcutta, India, interdes@satyam.net.in, II 216–219.

Giovannini Bernard, *Is it Necessary to Institutionalize Transdisciplinarity in Universities, and if It Is, How? M20*, DP MC 24, quai Ernest Ansermet, 1211 Genève, giovanni@sc2a.unige.ch, II 399–401.

Gisler Priska, *Obvious Facts or Invisible Histories? The Development of New Solar Technologies in an Transdisciplinary Perspective, D09*, Swiss Federal Institute of Technology, Dept. of History and Philosophy of Science, Schmelzbergstrasse 25, 8092 Zurich, gisler@wiss.huwi.ethz.ch, I 363–369.

Glauser Christoph, *New Information Technologies at the Edge of the 21st Century, D05*, MMS Media Monitoring Switzerland AG, Weltpoststrasse 20, 3000 Bern, glauser@mmsag.ch, I 189–190.

Gökalp Iskender, *Interdisciplinarité : une nécessité, tout simplement, I01*, CNRS-LCSR, 1C, avenue de la recherche scientifique, 45071 Orleans cedex 2, France, gokalp@cnrs-orleans.fr, I 515–517.

Goorhuis Henk, *Second Order Management for Emergent Problems in Nowadays Science and Society, D01*, University of Zurich, Fachstelle für Weiterbildung, Rämistrasse 74, 8001 Zurich, goorhuis@wb.unizh.ch, I 25–28.

Gotsch Nikolaus et al., *Polyproject "PRIMALP – Sustainable Agriculture and Forestry in the Alpine Region": Methodological Concept and Organisational

Needs of a Transdisciplinary Research Project at ETH, M13, Swiss Federal Institute of Technology, Agricultural Economics, ETH Zentrum, 8092 Zürich, nikolaus.gotsch@iaw.agrl.ethz.ch, II 280–285.

Gritsevich Inna, *Recent transdisciplinary Tendencies in Economies in Transition: From Academia to NGO and Consultancy, D04*, Center of Energy Efficiency, korp 4,54, Novocheremushkinsk str., 117418 Moscow, Russia, cenef@glas.apc.org, I 144–148.

Gros Henry, *The Consideration of the Entrepreneurial and Relational Dimensions for the Coaching of Transdisciplinary Research Teams, D04*, Innovation+ Management, Seidenstr.18, 8400 Winterthur, henry.gros@bluewin.ch, I 154–158.

Grütter Rolf et al., *Knowledge transfer in medical care: What information do doctors and patients need? M10*, Universität St. Gallen, Müller Friedberg-Strasse, 9000 St. Gallen, II 178–182.

Güldenzoph Wiebke et al., *Solid Waste – Transdisciplinary cooperative processes for sustainable solutions. Waste regulations on the potter's wheel, M14*, ETH Zentrum, 8092 Zürich, gueldenzoph@uns.umnw.ethz.ch, II 294–295.

Guthknecht Thomas, *Quality and Costs – How to marry two Enemies: Transdisciplinary Planning and Cost Control for Future Health Facilities, D06*, Itten + Brechbühl AG, Schoenburgstr. 19, 3000 Bern, tguthknecht@compuserve.com, I 244–247.

Habisch André et al., *Transdisciplinarity in Action: Promoting Corporate Citizenship as Investment in Social Capital, M17*, Universität Eichstätt, Center of Corporate Citizenship, Ostenstraße 26–28, 85072 Eichstätt, Germany, Andre.Habisch@ku-eichstaett.de, II 341–343.

Hadjieva Mila et al., *Transdisciplinary Practice for Implementation of Nonconvential Energy Sources in Bulgaria, M01*, Bulgarian Academy of Sciences, Central Laboratory of Solar Energy and New Energy Sources, 72 Tzarigradsko Schosse blvd., 1784 Sofia, Bulgaria, hadjieva@bas.bg, II 37–41.

Häfeli Ueli, *Strategies for Sustainable Transport – Experience from the Transisciplinary Research Project in the Framework of NFP 41 "Transport and Environment", D10*, University of Bern, Interdisciplinary Centre for General Ecology, Falkenplatz 16, 3012 Bern, haefeli@ikaoe.unibe.ch, I 384–387.

Haribabu E., *Cognitive Empathy as a Methodological Tool in Transdisciplinary Research: A Sociological Study of Biotechnology Research in India, M02*, University of Hyderabad, Department of Sociology, Hyderabad, India, ehbss@uohyd.ernet.in, II 52–53.

Harms Sylvia et al., *The Role of Users in Developing Sustainable Transport Practices – The Case of Car Sharing, D10*, Swiss Federal Institute for Environmental Science and Technology EAWAG, Überlandstr. 133, 8600 Dübendorf, sharms@eawag.ch, I 393–398.

Heeb Johannes et al., *Landscape Development in the Canton of Zurich: Reflecting a Pilot Project with Action Research, M13*, Seecon GmbH, Bahnhofstr. 2, 6110 Wollhusen, heebjohannes@pingnet.ch, II 260–264.

Hermanns Stengele Rita et al., *Groundwater-Remediation with Permeable Reactive Walls – a Transdiciplinary Approach to the Sustainable Use of the Most Important Natural Ressource, I02*, Swiss Federal Institute of Technology, Institute of Geotechnical Engineering, 8093 Zurich, hermanns@igt.baum.ethz.ch, I 590–595.

Herren Madelaine et al., *Studying Zoonoses at the Interface of Science and Society, M10*, Universität Zürich, Rämistr. 64, 8006 Zürich, mherren@hist.unizh.ch, II 187–191.

Hesske Stefan et al., *Transdisciplinary Environmental Education Example "Soil and Heavy Metals, I02*, UNS-HAT-ETH Zurich, 8092 Zurich, hesske@uns.umnw.ethz.ch, I 652–655.

Hiess Helmut, *Cultural Landscape 2020 – Future Images and Future Stories, D11*, Rosinak & Partner, Ziviltechniker , Gesellschaft m.b.H, Schloßgasse 11, 1050 Vienna, Austria, sekretariat@rosinak.co.at, I 416–419.

Hiremath B.N. et al., *Holistic Approach to Agricultural Technology Adoption: Insights from Semi-Arid Areas of India, D02*, Institute of Rural Management, Post Box - 60, Anand, India, kvr@fac.irm.ernet.in, I 51–55.

Hirsch Jemma Madeleine et al., *With Eight Points towards Sustainability: A Result of Transdisciplinarity, M05*, Geoprognos hirsch AG, Augustinergasse 25, 8001 Zurich, hirsch@swissonline.ch, II 101–104.

Hirsch Jemma Madeleine, *Managing Differences: Guidelines to perform Transdisciplinarity, M05*, Geoprognos hirsch AG, Augustinergasse 25, 8001 Zurich, hirsch@swissonline.ch, II 89–92.

Hoffelner Wolfgang et al., *Chances and Limitations of Solar Energy – A Transdisciplinary Approach, D03*, RWH consult GmbH, Buacherstrasse 10, 5452 Oberrohrdorf, rwh_whoffelner@compuserve.com, I 98–100.

Hoffmann Jürg, *Gobal Pollution Surveillance by Satellites, I02*, 3, rue St Blaise, 68990 Galfinque, France, j.hoffmann@wanadoo.fr, I 582–584.

Hoffmann-Riem Holger, *Basic Research on Transdisciplinary Problems: Lessons from a Case Ctudy on Environmental Problem Solving, D06*, Swiss Federal Institute of Technology, Institute of Terrestrial Ecology, Grabenstrasse 3, 8952 Schlieren, Hoffmann-Riem@ito.umnw.ethz.ch, I 220–223.

Holm Patricia, *Fishnet – a Transdisciplinary Project on the Decline of Fish Populations in Swiss River Systems, D12*, University of Bern, Interdisciplinary Centre for General Ecology/EAWAG, Falkenplatz 16, 3012 Bern/Dubendorf, patricia.holm@eawag.ch, I 446–449.

Howes Hugh R., *Synergies between Economic and Environmental Issues in South East England Opportunities Arising from New Constitutional Arrangements,*

M11, Environment Agency, Kings Meadow House, Kings Meadow Road, Reading, UK, hugh.howes@environment-agency.gov.uk, II 226–229.

Hugentobler Margrit, *Sustainably Urban Development as a Transdisciplinary Challenge, M11*, Swiss Federal Institute of Technology, Center for Housing and Sustainable Development, PO ETH Hönggerberg, 8093 Zurich, hugentobler@arch.ethz.ch, II 220–225.

Hurni Hans, *Addressing problems of global change through research, partnerships in an intercultural development context, D07*, Swiss Commission for Research Partnerships with Developing Countries (KFPE) of CASS, Centre for Development and Environment, Hallerstrasse 12, 3012 Bern, hurni@giub.unibe.ch, I 270–274.

Hvid Helge et al., *Dilemmas of Interdisciplinarity – A Case Story of Fission and Fusion in an Interdisciplinary Departement, M20*, Roskilde University, Department of Environment, Technology and Social Studies, PO 260, 4000 Roskilde, Denmark, hh@teksam.ruc.dk, II 386–390.

In't Veld Roel et al., *How to Optimise the Role of Science in Policy Making? Elucidations and Recommendations, I01*, Advisory Council for Research on Nature and the Environment, PO 5306, Rijswijk, The Netherlands, rmni@xs4all.nl, I 506–514.

Jabbar Mohammad et al., *From Component Technology to Integrated Resource Management: Evolution toward Transdisciplinary Research in a Project in Ethiopia, D07*, International Livestock Research Institute, PO 5689, Addis Ababa, Ethiopia, M.Jabbar@cgiar.org, I 275–279.

Jaeger Jochen et al., *Transdisciplinarity – just a Buzzword? Overcoming some Popular Objections to Transdisciplinary Research, D06*, Center of Technology Assessment in Baden-Württemberg, Industriestr. 5, 70565 Stuttgart, Germany, jochen.jaeger@ta-akademie.de, I 259–262.

Jäggi Olivier, *Searching for a Common Language, I02*, ECOFACT AG, Environment and Finance, Stampfenbachstrasse 42, 8006 Zurich, contact@ecofact.ch, I 617–618.

Jahn Thomas, *Status and Perspectives of Social-ecological Research in the Federal Republic of Germany, D02*, Institute for Social-ecological Research, Hamburger Allee 45, 60486 Frankfurt am Main, Germany, jahn@isoe.de, I 68–69.

Jansen J.L.A., *On Search for Ecojumps in Technology – From Future Visions to Technology Programms, D08*, Sustainable Technology Development – Knowledge Dissemination and Anchoring, PO 6063, 2600 JA Delft, Netherlands, jansen@dto.tno.nl, I 321–325.

Jeffrey, Paul et al., *Complex Systems Research as the Structuring of Cross-disciplinary Knowledge – Product Definition in the Aerospace Industry, I02*, Cranfield University, International Ecotechnology Research Centre & School of Water Sciences, MK43 0AL Cranfield, UK, p.j.jeffrey@cranfield.ac.uk, I 573–577.

Jeffrey Paul et al., *Reflections on the Practice of Cross-disciplinary Research; Towards an Understanding of the Mechanisms and Benefits of Collaboration, D05*, Cranfield University, International Ecotechnology Research Centre, MK43 0AL Cranfield, UK, p.j.jeffrey@cranfield.ac.uk, I 200–205.

Jenni Leo, *Transdisciplinary Research: The MGU-Research-Program, D03*, Koordinationsstelle MGU, Socinstrasse 59, 4002 Basel, leo.jenni@unibas.ch, I 85– 86.

Jenny Katharina et al., *The Indo-Swiss Collarboration in Biotechnology (ISCB), I02*, Institute of Biotechnology, Zürich, jenny@biotech.biol.ethz.ch, I 637–639.

Joos Walter et al., *Tools and Innovations for the Improvement of Acceptance in a Sustainable Waste Management System, M14*, Zurich University of Applied Sciences Winterthur, Dept. of Ecology, PO 805, 8401 Winterthur, crb@zhwin.ch, II 300–301.

Joss Simon, *Participatory Technology Assessment In The Public Sphere: Critical Reflections On The Relationship Between Institutionalised Forms Of Participation And The Public Sphere, M17*, University of Westminster, Centre for the Study of Democracy, 100 Park Village East, UK-London, UK, josss@wmin. ac.uk, II 331–334.

Kägi Wolfram, *The culture of sharing as a stabilisation mechanism of common property problems: An Anthropological-economic-psychological Research Project, D12*, University of St. Gallen, Institute for Economy and the Environment, Tigerbergstr. 2, 9000 St. Gallen, Wolfram.Kaegi@unisg.ch, I 472–476.

Kallis George et al., *Transdisciplinary Research in Urban Water Management: The Experience from a European Research Project, I02*, University of the Aegean, Dep. of Environmental Studies, Nikis 44, Athens, Greece, gkallis@ env.aegean.gr, I 657–658.

Kämpf Charlotte et al., *River Basin Management – a Transdisciplinary Approach: Experiences of transdisciplinarity in present research and plans to establish transdisciplinarity in a graduate programme, D12*, University of Karlsruhe, Institute of Water Resources Management, Hydraulic and Rural Engineering, 76128 Karlsruhe, charlotte.kaempf@bau-verm.uni-karlsruhe.de, I 467–471.

Kasanen Pirkko, *A Modest Success Story: LINKKI 2 Research Programme on Energy Conservation Decisions and Behavoiur, M01*, Työtehoseura (TTS) Institute (Work Efficiency Institute), PO, 00211 Helsinki, Finland, pirkko.kasanen@ pek.inet.fi, II 23–27.

Katzmann Werner, *Research on Cultivated Landscape (Kulturlandschaftsforschung) Transfer of Results into Education and Awareness-Raising. Some thoughts from Austria, M07*, Federal Institute for Public Health, Stubenring 6, 1010 Vienna, Austria, katzmann@oebig.at, II 145–148.

Kauffman Joanne M. et al., *The Alliance for Global Sustainability – A New, Integrated Model of Research for Better Decisions, Policies, and the Development of*

New Technologies, D08, Massachusetts Institute of Technology, 77 Massachusetts Ave., 02139 Cambridge, MA, USA, jmkauffm@mit.edu, I 336–340.

Kaufmann Ruth et al., *Instruments for Sustainable Development Results and Conclusions of a Transdisciplinary Project, D08*, University of Bern, Interdisciplinary Center for General Ecology, Falkenplatz 16, 3012 Bern, rkaufmann@ikaoe.unibe.ch, I 311–315.

Keitsch Martina M. et al., *Industrial Ecology Curriculum at Norwegian University of Science and Technology – An Interdisciplinary Approach, I02*, Norwegian University of Science and Technology, 7491 Trondheim, Norway, martina.keitsch@indecol.ntnu.no, I 600–604.

Keitsch Martina Maria, *Towards an Interdisciplinary Research in Technologies and Humanities – On the Concept of Industrial Ecology, M03*, Norwegian University of Science and Technology, 7491 Trondheim, Norway, martina.keitsch@indecol.ntnu.no, II 60–64.

Kesselring Annemarie, *Scare Resources in Nursing, M06*, Geschäftsstelle SBK, Choisystr. 1, 3001 Bern, kesselring.sbk@bluewin.ch, II 114–116.

KFPE, *Intercultural Learning – Mutual Learning in an Intercultural Context: Swiss Experiences of Transdisciplinary Research Partnership with Developing Countries, M18*, KFPE, Bärenplatz 2, 3011 Bern, kfpe@sanw.unibe.ch, II 346.

Kiefer Bernd et al., *Creating an Eco-label for Electricity in Switzerland – The Art of mediating between diverging interests, D09*, Trägerverein Ökostromlabel Schweiz, Schindlersteig 5, 8006 Zürich, Kieferpartners@access.ch, I 342–346.

Kiwi-Minsker L. et al., *Sustainable Chemical Technology for Improved Quality of Live, M09*, Ecole polytechnique fédérale, Institut de Génie Chimique, 1015 Lausanne, albert.renken@epfl.ch, II 166–167.

Klabbers Jan, *Enhancing the Effectiveness of Transdisciplinary Research in Handling Societal Risks: The Case of Global Climate Change, D06*, KMPC bv, Oostervelden 59, 6681 WR Bemmel, Netherlands, jklabb@antenna.nl, I 230–235.

Klaus Philipp, *Urban Quality, M11*, Güetliweg 1, Sonne, 8132 Hinteregg, klaus@smile.ch, II 198–201.

König Ilse et al., *Experiences of Transdisciplinarity within the Research Programme "Cultural Studies/Kulturwissenschaften", D08*, Austrian Federal Ministry of Science and Transport, Dept. for Social Sciences, Rosengasse 4, 1014 Vienna, Austria, ilse.koenig@bmwf.gv.at, I 304–308.

Kostecki Michel, *The Durable Use of Consumer Products: New Options for Business and Consumption, I02*, Université de Neuchâtel, The Enterprise Institute, Pierre-à-Mazel 7, 2000 Neuchâtel, 101767.153@compuserve.com, I 632–636.

Kovacevic Tea, *Impact Pathway Methodology in Estimating External Costs of Electricity Generation – Case Study "Croatia", M01*, Faculty of Electrical Engineering and Computing, Dpt. of Power Systems, Unska 3, Zagreb, Croatia, tea.kovacevic@fer.hr, II 31–36.

Krott Max, *Scientific Quality by Controlling: The Example of the interdisciplinary Programme "Austrian research program on cultural landscapes", D03*, Georg-August-Universität Göttingen, Institut für Forstpolitik und Naturschutz, Buesgenweg 5, 37077 Göttingen, Germany, mkrott@gwdg.de, I 115–119.

Kubasek Nancy K., *Green Taxes: An Alternative Way to Secure Environmental Protection, D09*, Bowling Green State University, Bowling Green, USA, nkubase@cba.bgsu.edu, I 358–362.

Kübler Markus et al., *Teaching Children Transdisciplinary Thinking: The Subject "Natur-Mensch-Mitwelt" (Nature-Man-Environment) as a New Approach to Transdisciplinary Learning in the Primary School, D05*, Teachers Training College, Schlüsselmattenweg 23, 3700 Spiez, mkuebler@bluemail.ch, I 211–217.

Kulikauskas Paulius, *Transdisciplinarity in Planning of Sustainable Urban Revitalisation, M11*, Byfornyelsesselskabet Danmark, Urban renewal company, Studiestraede 50, 1554 Copenhagen, Denmark, pauliusk@counsellor.com, II 230–235.

Künzi Erwin, *Transdisciplinarity in the Laikipia Research Programme – Research and Co-operation for Sustainable Development in a Semi-Arid Regional Context in Kenya, M18*, University of Bern, Interdisciplinary Centre for General Ecology, Falkenplatz 16, 3012 Bern, e.kuenzi@magnet.at, II 363–368.

Künzli Christine et al., *Skills or Content? How to Prepare Students for Interdisciplinary and Transdisciplinary Research and Practice, I02*, University of Bern, Interdisciplinary Centre for General Ecology, Falkenplatz 16, 3012 Bern, christine.kuenzli@ikaoe.unibe.ch, I 619.

Küry Daniel, *Analysing the Perception of River Restorations: Contributions to New Conservation Strategies, D03*, Life Science AG, Greifengasse 7, 4058 Basel, kueryd@ubaclu.unibas.ch, I 88–89.

Kvarda Werner, *New Bridges to Learn: Models of Good Practice – The "Bridge Danube Project", M11*, Univ. of Agricultural Sciences Vienna, Institute for Landscape Architecture and Landscape Management, 1190 Vienna, Austria, freiraum@mail.boku.ac.at, II 211–215.

Kyrtsis Alexandros-Andreas, *Transdisciplinarity and Innovation in Banking, M03*, University of Athens, Department of Economics, 8 Pesmazoglou Street, 10559 Athens, Greece, kyrtsis@compulink.gr, II 65–69.

Lawrence Roderick J., *Overcoming Obstacles to Transdisciplinary Research and Policy Implementation: Building Partnerships, M20*, University of Geneva, Centre for Human Ecology & Environmental Sciences, Bd Carl-Vogt 102, 1211 Geneva, lawrence@uni2a.unige.ch, II 396–398.

Lechner Robert, *Culture, Landscape, Development in Alpine Regions of Western Austria, M13*, Austrian Institute for Applied Ecology, Seidengasse 13, 1070 Vienna, Austria, oekoinstitut.plan@ecology.at, II 274–278.

Lenz Roman, *Project Overview European-Canadian-Curriculum on Environmental Informatics (ECCEI), M09*, Nuertingen University for Applied Sciences,

Schelmenwasen 4–8, 72622 Nuertingen, Germany, lenzr@fh-nuertingen.de, II 157–159.

Lima Jr Fontes Orlando et al., *Civil Engineer Environmental Literacy: A Transdisciplinary Proposal, I02*, UNICAMP – State University of Campinas, Caixa Postal 6021, 13.083-970 Campinas, Brasil, emilia@fec.unicamp.br, I 567–572.

Limpert Eckart et al., *Life is Log-Normal: Keys and Clues to understand Patterns of Multiplicative Interactions from the Disciplinary to the Transdisciplinary Level, D01*, Swiss Federal Institute of Technology, Institute of Plant Sciences, Eschlikon 33/PO 185, 8315 Lindau, eckhard.limpert@ipw.agrl.ethz.ch, I 20–24.

Limpert Eckart et al., *Life is Log-Normal: On the Charms of Statistics for Society, I01*, Swiss Federal Institute of Technology, Institute of Plant Sciences, Eschlikon 33/PO 185, 8315 Lindau, eckhard.limpert@ipw.agrl.ethz.ch, I 518–522.

Loibl Marie Céline, *Group-Dynamics in Transdisciplinary Research, D04*, Austrian Institute for Applied Ecology, Seidengasse 13, 1070 Vienna, Austria, oekoinstitut.plan@ecology.at, I 131–134.

Lukesch Robert et al., *The Green Leaves of Life's Golden Tree: The project "Research in Public" – Suburbanisation Processes in Rural Areas, D11*, ÖAR-Regionalberatung GmbH, Hierzenriegl 55, 8350 Fehring, Austria, lukesch@eunet.at, I 431–435.

Lunca Marilena, *Crossing and Unifying Disciplinary Languages, I01*, Laan van Chartoise 108, 3552 EX Utrecht, The Netherlands, river@river.tmfweb.nl, I 505.

Maier Simone et al., *Constraints & Options – Sustainable Development within the Need-Field of Nutrition, D02*, University of St. Gallen, Institute for Economy and the Environment, Tigerbergstr. 2, 9000 St. Gallen, Simone.Maier@idheap.unil.ch, I 46–50.

Markard Jochen et al., *Green Power in liberalised electricity markets: Market experiences and policy implications as examples for transdisciplinary research, D09*, Swiss Federal Institute for Environmental Science and Technology, Seestrasse 79, 6047 Kastanienbaum, jochen.markard@eawag.ch, I 352–357.

Martínez Sylvia et al., *Coordination as a Prerequisite for Successful Transdisciplinary Research Collaboration, I01*, MCO Biodiversity, Swiss Priority Program Environment (SPPE), Universitiy of Basel and Zurich, Schoenbeinstrasse 6, 4056 Basel, mco@ubaclu.unibas.ch, I 492–494.

Mathys Renata, *Managing the Knowledge Potential between Industry, Society and Academia – A Case Example of the School of Engineering Burgdorf, M09*, Applied University of Bern, School of Engineering Burgdorf, Pestalozzistrasse 20, 3400 Burgdorf, renata.mathys@hta-bu.bfh.ch, II 156.

Mauch Corine, *The Role of Science in Local Agenda 21 Processes, M12*, topos Marti & Müller, Idastr. 24, 8003 Zurich, topos@access.ch, II 242.

Mebratu Desta, *Transdisciplinarity and the Developing World, D07*, UNECA/RCID, PO 8412, Addis Ababa, Ethiopia, dmebratu@hotmail.com, I 289–294.

Messerli Peter, *The Dilemma between Development and Conservation on the Eastern Escarpment of Madagascar – Experiences of Problem-oriented and Applied Research in a Developing Country, M18*, University of Bern, Centre for Development and Environment, Projet BEMA, B.P. 4052, Antananarivo, Madagascar, messerli@dts.mg, II 352–357.

Michelsen Gerd et al., *Looking Forward: A Contribution to "Sustainable Developmen" in Universities, D05*, Universität Lüneburg, Scharnhontstrasse 1, 21335 Lüneburg, michelsen@uni-lueneburg.de, I 176–178.

Mirenowicz Jacques, *Consensus Conference and Transdisciplinarity: A Tale of Great Expectations, M17*, Institut pour la Communication et l'analyse des sciences et des technologies, Place Notre-Dame, 8, 1700 Fribourg, Jacques. Mirenovicz@icast.org, II 315–317.

Mogalle Marc, *The Need-Field Approach: Illustrated by the Example of the Integrated Project Society I/ SSP Environment "A Sustainable Switzerland in the International Context", D02*, University of St. Gallen, Institute for Economy and the Environment, Tigerbergstr. 2, 9000 St. Gallen, Marc.Mogalle@unisg.ch, I 70–74.

Moritsuka Hideto, *Carbon Dioxide Recovery Technology for China, I02*, Central Research Institute of Electric Power Industry (CRIEPI), Yokosuka Research Laboratory, Address: 2-6-1 Nagasaka, Japan - 240-0196, moritsuk@criepi.denken.or.jp, I 565–566.

Moser Karin S. et al., *Taking Actors' Perspectives seriously: Who's Knowledge and what is managed? Knowledge Management in a Transdisciplinary Perspective, I01*, University of Zurich, Dept. of Psychology, Plattenstrasse 14, 8032 Zurich, kmoser@sozpsy.unizh.ch, I 534–538.

Naveh Zev, *Transdisciplinary Challenges for Regional Sustainable Development Toward the Post-Industrial Information Society, D02*, Faculty of Agricultural Engineering, Technion, 32000 Haifa, Israel, znave@hotmail.com, I 78–82.

Nentwich Michael, *The Role of Particpatory Technology Assessment in Policy-making, M17*, Institute of Technology Assessment, Inselgasse 1, 3003 Bern, II 335–339.

Neumann-Held Eva et al., *Philosophy and Developmental Genetics A Case Study for Transdisciplinary Research, D03*, Europäische Akademie GmbH, Wilhelmstr. 56, 53474 Bad Neuenahr-Ahrweiler, Germany, eva.neumann-held@dlr.de, I 90–92.

Nunes Paulo Augusto, *Using CV to Measure Environmental Protection Benefits: Is there any Warmglow in the Stated WTP Responses? I02*, Catholic University Leuven, Centre for Economic Studies, 69 Naamsestraat, 3000 Leuven, Belgium, pnunes@econ.vu.nl, I 663.

Oguttu Wogenga'h H. et al., *Point Source Pollution from the Catchment Area into Jinjas Urban Wetlands, Uganda, D07*, Fisheries Research Institute, P.O. Box 343 Jinja, Uganda, firi@infocom.co.ug, I 282–283.

Oswald Jennifer E. et al., *Organizing Mutual Learning Processes between Science and Society: Lessons learned in 6 Years Experiences with the ETH-UNS Case Study, D04*, Swiss Federal Institute of Technology, Natural & Social Science Interface, ETH Zentrum HAD, Haldenbachstrasse 44, 8092 Zurich, fsbuero@uns.umnw.ethz.ch, I 135–140.

Ouboter Stefan et al., *Negotiating towards a desired environment, D12*, Centre for Soil Quality Management and Knowledge Transfer, P.O. Box 420, 2800 AK Gouda, Netherlands, skb@cur.nl, I 454–458.

Penkova Nathalia V., *Regional Aspects of Water-Economic Development in a Transdisciplinary Dimension: Unique Units and General Methodology, D12*, State Hydrological Institute, 23 Second line, 199053 St.-Petersburg, Russia, penkov@peterlink.ru, I 461–466.

Perrig-Chiello Pasqualina et al., *Inter- and Transdisciplinary Activities at Swiss Universities: The Development and Restructuring of a Database, I02*, Institut Universitaire Kurt Bösch, PO 4176, 1950 Sion, pasqualina.perrigchiello@ikb.vsnet.ch, I 605–609.

Perrig-Chiello Pasqualina et al., *Transdisciplinary Research on Ageing: Theory and Practice, I02*, Institut Universitaire Kurt Bösch, PO 4176, 1950 Sion, pasqualina.perrigchiello@ikb.vsnet.ch, I 659–662.

Petersen Holger et al., *NGOs: Promoters of Sustainable Entrepreneurship? I01*, University of Lüneburg, Chair of Corporate Environmental Management, Scharnhorststraße 1, 21335 Lüneburg, Germany, hpetersen@uni-lueneburg.de, I 523–527.

Ping Xiao, *The Impact of Lifestyle and Consumption Preference Changes on Land Use in China: Facts, Mechanisms and a Framework for Further Research, D11*, Wuhan East Lake High Technology Group CO.,LTD., 11-11 Wan Song Yuan Road, Jiang Han District, P.R. 430022 Wuhan, China, xiaoping16@hotmail.com, I 436–443.

Pivot Agnes et al., *Natures Sciences Sociétés (NSS), The outcome of twenty years of interdisciplinary and inter-institutional research collaboration among the life sciences, earth sciences and social sciences, D02*, NSS/CNRS, Université de Paris X, 200 Avenue de la République, 92001 Nanterre, France, apivot@u-paris10.fr, I 56–61.

Plattner Rolf, *Living in Basel – a project course cycle within the MGU-education program, D03*, Plattner Schulz Partner, Steinenring 10, 4051 Basel, pspag@pspag.ch, I 101–103.

Pohl Christian, *Inter- and Transdisciplinary Research Methods: What Problems They Solve and How They Tackle Them, D01*, Swiss Federal Institute of Technology, SAGUF Geschäftsstelle, ETH Zentrum HAD, Haldenbachstrasse 44, 8092 Zurich, saguf@umnw.ethz.ch, I 18–19.

Pokorny Doris et al., *Sustainability through Transdisciplinarity? The Biosphere Reserve Concept as Opportunity and Challenge, D11*, Biosphärenreservat Rhön, Marktstr. 41, 97656 Obereisbach, Germany, brrhoenvsbay@swin.de, I 425–430.

Popow Gabriel, *How Farmers are Educated in the Areas of Nature and Landscape, M13*, LIB Strickhof, Eschikon, 8315 Lindau, II 279.

Rais Mohammad, *Integrating Transdisciplinarity in Assessing Sustainability: A Dialogue between Research Framework and Farmers Perception: Some Experiences of Sloping Lands from Southeast and South Asia, D11*, National Institute of Science Technology and Development Studies, Resource Planning and Utilization for Regional Development, Dr K.S. Krishnan Road, New Dehli, India, mohammad_rais@hotmail.com, I 420–424.

Raju K.V. et al., *Science in Society: Social and Ecological Ethics, D04*, Institute of Rural Management, PO 60, Anand, India, kvr@fac.irm.ernet.in, I 149–153.

Rauschmayer Felix, *Legitimisation of Decision Making in the Context of Sustainable Development, D06*, Institut für Philosophie, Burgstr. 21, 04109 Leipzig, Germany, rauschma@rz.uni-leipzig.de, I 236–238.

Rege Colet Nicole, *Evaluating Interdisciplinary Teaching, D03*, Université de Genève, Adjointe au Rectorat à la formation et à l'Evaluation, 24, rue Général Dufour, 1211 Geneva, Nicole.RegeColet@rectorat.unige.ch, I 120–122.

Reichl Franz, *Facilitated Open Distance Learning for Continuing Engineering Education, M07*, Vienna University of Technology, Gusshausstrasse 28, 1040 Vienna, Austria, Franz.Reichl@tuwien.ac.at, II 138–142.

Reinhardt Ernst, *Eco-DriveTM in Switzerland – A Success Story of Energy 2000: Improving Fuel Efficiency in Road Transport, D10*, ecoprocess, Leonhardshalde 21, 8001 Zurich, ernst.reinhardt@ecoprocess.ch, I 380–383.

Ritz Christoph et al., *From Knowledge to Action – Transdisciplinarity in Climate Research, D12*, ProClim, Bärenplatz 2, 3011 Bern, ritz@sanw.unibe.ch, I 450–453.

Robledo Carmenza et al., *Sustainable Development of the Tropical Forest in Colombia, I02*, Swiss Federal Institute of Technology, Natural & Social Science Interface, ETH Zentrum HAD, Haldenbachstrasse 44, 8092 Zurich, sell@uns.umnw.ethz.ch, I 625–626.

Rubin de Celis Emma et al., *Transdisciplinarity in practice, Lessons from an international action-research, and development project, M06*, Institute of Tropical Medicine, Nutrition Unit, 2000 Antwerpen, Belgium, plefevre@itg.be, II 124–128.

Sauvain Paul, *Sustainable Skitourism in the Age of Globalisation: Present Challenges, Questions and new Concepts, M05*, BEREG, 1934 Bruson, serec.brus@gve.ch, II 98–100.

Schenkel Walter, *Transdisciplinary Research and Co-operation in Sustainable Urban Policy, M11*, Muri&Partner, Limmatquai 1, 8001 Zurich, schenkel@pwi.unizh.ch, II 236–240.

Schenler Warren W., *SESAMS (Strategic Electric Sector Assessment Methodology under Sustainability), D09*, Swiss Federal Institute of Technology, Institute for Energy Technology, Weinbergstrasse 11, 8001 Zurich, schenler@pst.iet.mavt.ethz.ch, I 373.

Scheringer Martin et al., *Transdisciplinarity and Holism – How are Different Disciplines Connected in Environmental Research? D01*, Swiss Federal Institute of Technology, Laboratory of Chemical Engineering, ETH-Zentrum, 8092 Zurich, scheringer@tech.chem.ethz.ch, I 35–37.

Scholz Roland W., *Introduction to the Mutual Learning Session Workbook of the Transdisciplinarity 2000 Conference, M00*, Swiss Federal Institute of Technology (ETH), Chair of Environmental Sciences – Natural and Social Science Interface, Haldenbachstr. 44, 8092 Zurich, scholz@uns.umnw.ethz.ch, II 10–12.

Scholz Roland W., *Mutual Learning as a Basic Principle of Transdisciplinarity, M00*, Swiss Federal Institute of Technology (ETH), Chair of Environmental Sciences – Natural and Social Science Interface, Haldenbachstr. 44, 8092 Zurich, scholz@uns.umnw.ethz.ch, II 13–17.

Schönlaub Hans P., *Public Understanding of Geosciences – a Transdisciplinary Approach, D05*, Geological Survey of Austria Vienna, Austria, schhp@cc.geolba.ac.at, I 198–199.

Schübel Hubert R., *Psychological Monitoring and Team-Coaching for the Improvement of Inter-/Transdiciplinary Processes, D04*, Consultant for Organizational Psychology, Elisabethenstr. 30, 70197 Stuttgart, Germany, Schuebelhr@aol.com, I 141–143.

Schulz Klaus-Peter et al., *Study to Work – An Interdisciplinary Access to the Field of Innovation Management in University Education, D06*, TU Chemnitz, Lehrstuhl fuer Management des technischen Wandels und Personalentwicklung, BWL IX, 09107 Chemnitz, Germany, schulzkp@wirt.wirtschaft.tu-chemnitz.de, I 248–253.

Schüpbach Hans, *The Agricultural Knowledge System in Switzerland. Institutional Innovations for Joint Problem Solving among Practice and Science, M13*, Agricultural Extension Center, Eschikon, 8315 Lindau, hans.schuepbach@lbl.ch, II 265–268.

Schwarz Astrid E., *The "Lake" as a Mirror of Cultural Identity – Ecological Pictures of Nature in the Socio-political Context, D01*, TU München, Lehrstuhl für Landschaftsökologie, Dorfstrasse 35, 85350 Freising, Germany, Astrid.Schwarz@spectraweb.ch, I 32–34.

Schweizer Peter, *Computer Aided Invention (CAI). Will CAI become a new tool for lone fighters to raise their survival rate? I01*, Freiestrasse 131, 8032 Zurich, schweizer.peter@active.ch, I 501–504.

Schweizer Peter, *We do not want to solve problems. We want to be confirmed in our ideas! D01*, Freiestrasse 131, 8032 Zurich, schweizer.peter@active.ch, I 39–43.

Semadeni Marco, *Moving from Risk to Action: A conceptual risk handling model, D06*, Swiss Federal Institute of Technology, Natural & Social Science Interface, ETH Zentrum HAD, Haldenbachstrasse 44, 8092 Zurich, semadeni@uns.umnw.ethz.ch, I 239–243.

Sorg Jean-Pierre, *Transdisciplinarity: a Small-Scale Comparative Analysis of the Research Activities in two Forest Management Projects in Madagascar, D07*, Swiss Federal Institute of Technology, Waldbau, Rämistrasse 101, 8092 Zurich, sorg@waho.ethz.ch, I 284–288.

Sotoudeh Ariane, *Sustainable development in Switzerland, M12*, BUWAL, PO, 3003 Bern, ariane.sotoudeh@buwal.admin.ch, II 256–257.

Spaapen Jack B. et al., *The Evaluation of Transdisciplinary Research, D03*, sci_Quest, research agency for S&T Policy, Herculesstraat 43, 1076 Amsterdam, jbspaa@xs4all.nl, I 111–114.

Stalder Ueli, *The Mutual Learning Session "Sustainable Tourism": Situation, Objectives, Programme, M05*, University of Bern, Economic Geography and Regional Studies, Hallerstrasse 12, 3012 Bern, stalder@giub.unibe.ch, II 82–84.

Steffany Frank, *Programme and Research Results of Special Research Initiative 419 Supported by the "Deutsche Forschungsgemeinschaf", D08*, University of Cologne, Institute of Geophysics und Meteorology, Kerpener Straße 13, 50937 Cologne, Germany, steffany@meteo.Uni-Koeln.de, I 309–310.

Steurer Johann, *Information transfer from science into general practice, M10*, Universitätsspital, Medizinische Polyklinik, Rämistrasse 100, 8091 Zurich, johann.steurer@dim.usz.ch, II 176–177.

Stuhler Elmar A., *Sustainable Development – A Challenge To The Universities, I01*, Technische Universität München, Institut für Wirtschaft und Soz. Wissenschaft, Munchen, Germany, stuhler@pollux.weihenstephan.de, I 610–614.

Suhr Nelson Julie, *Transdisciplinary Decision Making Processes: A Methodology for the Analysis of Qualitative, Quantitative and Economic Data, D04*, University of Utah, Department of Economics, Salt Lake City, USA, jnelson1@ uswest.net, I 159–163.

Sundin Nils-Göran, *Science and the Arts: Trespassing the last Taboo towards a Phenomenology of Interpretation in Performance, M19*, Radmansgatan 3, 11425 Stockholm, nils-goran.sundin@mailcity.com, II 374–379.

Tanner Marcel et al., *Management of Wastewater Use in Small-Scale Urban Agriculture: From Research to Action, M18*, Swiss Tropical Institute, Support Centre for International Health, Socinstr. 57, 4002 Basel, tanner@ubaclu.unibas.ch, II 347–351.

Taroni Franco et al., *Law, Forensic Science and Statistics – the Need for a Collaboration, D08*, Université de Lausanne, Institut de medecine legale, 21 rue du Bugnon, 1005 Lausanne, Franco.Taroni@inst.hospvd.ch, I 316–320.

Toussaint Verena et al., *Approaches to Sustainable Agricultural Production: Models for North-eastern Germany – Management of a Transdisciplinary Research Project, M13*, Zentrum für Agrarlandschafts- und Landnutzungsforschung, Institut für Sozioökonomie, Eberswalder Str. 84, 15374 Müncheberg, Germany, vtoussaint@zalf.de, II 269–273.

Truffer Bernhard et al., *The Management of "Strategic Niches" for promoting Sustainable Transport Innovations, D10*, Swiss Federal Institute for Environmental Science and Technology, 6047 Kastanienbaum, Seestrasse 79, truffer@eawag.ch, I 388–392.

Truffer Bernhard, *Setting the Stage for a new Kind of Research – The social Construction of Green Electricity Standards in Switzerland, D09*, Swiss Federal Institute for Environmental Science and Technology, Seestrasse 79, 6047 Kastanienbaum, truffer@eawag.ch, I 374–378.

Ulli-Beer Silvia, *A Transdisciplinary System Dynamics Approach to Responsible Environmental Behaviour: MODEL-REB, D04*, University of Bern, Interdisciplinary Centre for General Ecology, Falkenplatz 16, 3012 Bern, ulli@ikaoe.unibe.ch, I 126–130.

Ulrich Markus et al., *A Method to Communicate Scientific Results to People Outside Science – the Example of "Corebifa – Simulation Game on Sustainable Development and Nutrition", I02*, UCS Ulrich Creative Simulations, Simulation & Gaming/Policy Exercises for Education and Conflict Resolution, Blaufahnenstrasse 14, 8001 Zurich, ucs@access.ch, I 560–564.

Ulrich Markus, *Strategies for Sustainable Development in the Need Field Nutrition – An Innovative Approach Using the Policy Exercise "Simulation Game on Sustainable Nutrition", M04*, UCS Ulrich Creative Simulations, Simulation & Gaming/Policy Exercises for Education and Conflict Resolution, Blaufahnenstrasse 14, 8001 Zurich, ucs@access.ch, II 72–77.

Vahtar Marta, *Transdisciplinarity Requires Proactive Culture: Experiences of Transdisciplinarity in Slovenia from an Academic Research and from a NGO Activist's Perspective, I02*, Institute for Integral Development and Environment, Savska 5, 1230 Domzale, Slovenia, marta.vahtar@guest.arnes.si, I 656.

Van de Kerkhof Marleen et al., *The Dutch COOL-Project: Bringing Transdisciplinarity into Practice? D08*, Institute for Environmental Studies, Boelelaan 1115, 1081 HV Amsterdam, marleen.van.de.kerkhof@ivm.vu.nl, I 296–300.

Van Veen Johan, *Improvement of soil quality management by the stimulation of transdisciplinary projects, I02*, Centre for Soil Quality Management and Knowledge Transfer, P.O. Box 420, 2800 AK Gouda, Netherlands, skb@cur.nl, I 596–599.

Vermeulen W.J.V. et al., *Organising "CO2-Reduction Options Markets": Bridging Multiple Chasms within the Science-Policy-Society Triangle, D12*, TNO Strategy, Technology and Policy, Delft, Netherlands, W.Vermeulen@geog.uu.nl, I 459–460.

Villa Alessandro E.P. et al., *Transdisciplinary Approach to Scientific Data Analysis Through Internet, I01*, Université de Lausanne, Laboratoire de Neuroheuristique, Institut de Physiologie, 7, Rue du Bugnon, 1005 Lausanne, Alessandro.Villa@iphysiol.unil.ch, I 550–555.

Von Reding Walter, *Virtual Teams in the Network Society or Mapping the Invisible Workplace (Theoretical part), M10*, MBA Network Economics, Rossberg, 6422 Steinen, waltervonreding@csi.com, II 192–196.

Wackers Ger et al., *Transdisciplinarity and Transportation, D10*, University of Maastricht, Herbenusstraat 51, 6211 Maastricht, Netherlands, veldro@ilimburg.nl, I 399–403.

Wäger Patrick et al., *A Dynamic Model for Decision-Making in Plastics Waste Management, M14*, Swiss Federal Laboratories for Materials Testing and Research, Ecology, Lerchenfeldstr. 5, 9014 St. Gallen, patrick.waeger@empa.ch, II 296–299.

Wahed Najib Abdul, *Transdisciplinarity Models of the Golden Ages: Lessons and Proposals, D01*, University of Aleppo, Faculty of Mechanical Engineering, Aleppo, Syria, najibaw@cyberia.net.lb, I 38.

Wanschura Bettina et al., *Public Relations for and Marketing of Scientific Research Results, D05*, Büro Plansinn, Schleifmühlgasse 1a/14, 1140 Vienna, plansinn@teleweb.at, I 196–197.

Weber Lukas, *Energy Analysis and Theories of Action: Energy-Relevant Decisions Within Office Buildings, D09*, Swiss Federal Institute of Technology, Centre for Energy Policy and Economics CEPE, ETH Zentrum, ETL G 22, 8092 Zurich, lweber@pauke.ethz.ch, I 347–351.

Weber Olaf, *Sustainable Banking -Relationship between banks, companies, society and environment, M03*, Swiss Federal Institute of Technology, Natural & Social Science Interface, ETH Zentrum HAD, Haldenbachstrasse 44, 8092 Zurich, weber@uns.umnw.ethz.ch, II 56–59.

Wehrli Bernhard et al., *Scientific Research for Defining the sustainable Use of Hydropower, D09*, Swiss Federal Institute for Environmental Science and Technology, Seestrasse 79, 6047 Kastanienbaum, bernhard.wehrli@eawag.ch, I 370–372.

Weiss Martina, *Dialogue between Sciences, Literature and Fine Arts – in Search of Quality Criteria, M19*, Collegium Helvetivum, STW B5.2, ETH-Zentrum, 8092 Zurich, weiss@collegium.ethz.ch, II 370–371.

Wils Jean-Pierre, *The Rationing System of Health Care Services in the Netherlands, M06*, University of Zurich/Interdisziplinäres Institut für Ethik im Gesundheitswesen, Gloriastrasse 18, 8028 Zurich, info@dialog-ethik.ch, II 117–118.

Winistörfer Herbert et al., *Tools needed for successful transdisciplinary research: the Social Compatibility Analysis, I01*, Zurich University of Applied Sciences Winterthur, Dept. of Ecology, PO 805, 8401 Winterthur, Win@zhwin.ch, I 539–543.

Wyss Kaspar et al., *How to Achieve Effectively the Management of Urban Environmental Issues? – Environmental and Social Mobilisation in N'Djamena, Chad and Dakar, Senegal, M11*, Swiss Tropical Institute, Socinstr. 57, 4002 Basel, wyssk@ubaclu.unibas.ch, II 202–205.

Yonkeu Samuel et al., *The Transdisciplinary Approach in the Integration of Environmental Sciences in the Training Programme of the inter-state College for Rural Engineering, Ouagadougou, Burkina Faso, M09*, Inter-State School of Engineers of Rural Equipment, (EIER) 03, P.O. Box, 7023 Ouagadougou, Burkina Faso, yonkeu@eier.univ-ouaga.bf, II 168–173.

Yun Guichun et al., *Wastewater Reclamation and Recycling to Resolve Water Shortage Problem in China, M01*, Technology Tsinghua University, Institute for Nuclear Energy, Beijing, China, tianbao@inet.tsinghua.edu.cn, II 42–43.

Zacher Lech W., *Possible Status of Transdisciplinarity, M19*, Acad. of Entrepreneurship and Management, Hagiellonska 59, 03-301 Warsaw, Poland, Lzacher@wspiz.edu.pl, II 372–373.

Zoller Uri, *HOCS in the STES Context – an Imperative for the Disciplinary-Transdisciplinary Paradigm Shift, M07*, Haifa University Oranim, Dept. of Science, Educ.-Chemistry, Kiryat Tivon 36006, Haifa, Israel, uriz@research.haifa.ac.il, II 143–144.

Abbreviations and Acronyms

ABB	Asea Brown Boveri Corporation
ETH	Swiss Federal Institute of Technology
ETH-UNS	Swiss Federal Institution of Technology, Natural and Social Science Interface
EU	European Union
IDRC	International Development Research Center, Ottawa, Canada
IKAO	Interdisciplinary Center for General Ecology, University of Berne
ISOE	Institute for Social-Ecological Research, Frankfurt am Main
MGU	Man-Society-Environment Program, University of Basel
MISTRA	Swedish Foundation for Strategic Environmental Research
MIT	Massachusetts Institute of Technology
MOST	Management of Social Transformation program, UNESCO
NGO's	Non-governmental organizations
NSF	National Science Foundation
OECD	Organization for Economic Cooperation and Development
R&D	Research and Development
RTD	Research, Technology and Development
SAFEL	Swiss Agency for Forest, Environment and Landscape
SAGUF	Swiss Academic Society of Environmental Research and Ecology
SPPE	Swiss Priority Program Environment
TA	Technology Assessment
WWF	World Wildlife Fund

Code for Conference Sessions

D	Dialogue Sessions
I	Idea Market
M	Mutual Learning Sessions

MIX
Papier aus verantwortungsvollen Quellen
Paper from responsible sources
FSC® C105338

Printed by Books on Demand, Germany